CHEMISTRY AND ENERGY

I0050997

CHEMISTRY AND ENERGY

A Consumer Chemistry Course

First Edition

Terrence Lee

Middle Tennessee State University

cognella®

SAN DIEGO

Bassim Hamadeh, CEO and Publisher
Spencer Szwalbenest, Associate Acquisitions Editor
Skyler Van Valkenburgh, Project Editor
Susana Christie, Senior Developmental Editor
Samantha Hansen, Production Editor
Emely Villavicencio, Senior Graphic Designer
Kara Tatum, Licensing Coordinator
Natalie Piccotti, Director of Marketing
Kassie Graves, Senior Vice President, Editorial
Alia Bales, Director, Project Editorial and Production

Copyright © 2025 by or Cognella, Inc. All rights reserved. No part of this publication may be reprinted, reproduced, transmitted, or utilized in any form or by any electronic, mechanical, or other means, now known or hereafter invented, including photocopying, microfilming, and recording, or in any information retrieval system without the written permission of Cognella, Inc. For inquiries regarding permissions, translations, foreign rights, audio rights, and any other forms of reproduction, please contact the Cognella Licensing Department at rights@cognella.com.

Trademark Notice: Product or corporate names may be trademarks or registered trademarks and are used only for identification and explanation without intent to infringe.

Cover image copyright © 2018 iStockphoto LP/coffeekai.

Printed in the United States of America.

cognella® | ACADEMIC PUBLISHING
320 South Cedros Ave., Ste. 400, Solana Beach, CA 92075

Contents

Preface

I have taught a college-level consumer chemistry course, focusing on chemistry and energy, since 2014. While there are several textbooks on consumer chemistry, none of them focus on energy and chemistry. Rather than have my students buy a textbook with limited coverage of energy, I chose to teach the course relying on notes covering the topics I thought important.

Not having a textbook has its advantages. It saves the students money; it allows the instructor flexibility to incorporate new material, or to delete material that becomes dated or irrelevant; and it allows the course to be regularly updated to incorporate the most recent developments. The disadvantages are just as important. Without a textbook, students are dependent on lecture attendance as their primary means of getting the information. They don't have a convenient, readily available resource to use when studying. If they are being tutored, their tutor might not be completely familiar with the material. If the instructor is absent, the substitute might not be familiar with the material.

I decided to write a textbook for my specific course when I was approached by Cognella about the possibility of writing a textbook for consumer chemistry. We quickly came to an agreement about the scope of the work, and you are holding the result in your hands.

I want to thank the wonderful people at Cognella who made this book possible, but especially Michelle Piehl, Senior Project Editor and now Production Specialist. This is a great team of publishing professionals whose help was vital in producing this book.

Introduction

"Energy is the lifeblood of modern technologically advanced societies."

I don't know who first said this, but it wasn't me, so I'll put it in quotes.

Energy is critical to every economic activity and noneconomic activity of our society. Energy policy—how we obtain, use, and regulate energy—directly impacts our lives and the lives of our children and grandchildren, and their grandchildren. Energy decisions made today will affect our country for generations.

Energy is much too important to the survival of society to be left solely to the politicians, the technologists, the industrialists, the "aristocrats." The average person may not be allowed to make energy policy or directly influence energy decisions, but it is absolutely vital that people have some basic knowledge in order to understand the energy choices and decisions being made on their behalf.

This book is meant to give the general reader a background to understand energy from historical, scientific, and societal viewpoints. The first six chapters provide enough background in physics and chemistry for the reader to make sense of the last five chapters. The last five chapters provide a summary overview of current (2020–2022) energy resources and usage in the United States.

For several years, I've taught a general education, college-level course for nonscience majors using this material. Students have responded positively to this course, but I am always looking for improvements. If you have suggestions for improving the book, I'd be glad to consider them.

Dr. Terrence A. Lee
MTSU Box 68
Middle Tennessee State University
Murfreesboro, TN 37132

First Energy Sources for Mechanical Work

> We may consequently regard it as certain that, neither by natural agencies of inanimate matter, nor by the operations arbitrarily effected by animated Creatures, can there be any change produced in the amount of mechanical energy in the Universe.
>
> William Thomas Thomson, Baron Kelvin of Largs

The United States, like all technologically advanced countries, consumes vast amounts of energy. In 2020, the U.S. used 92,940,000,000,000,000 **BTU** (British Thermal Units) of energy—this is 309,800,000 BTU per person, or about 35,365 BTU/hour per person. (For comparison, an average person sitting at home produces about 356 BTU/hour.) Where does this energy come from? How is it produced? What is it used for? What are the benefits and what are the problems involved? To answer these questions, we need to start at the beginning.

Learning Objectives

This chapter will help students

- Gain an understanding of the earliest sources of mechanical work
- Develop an understanding of the limitations of early sources of mechanical work

BTU: the amount of heat needed to raise the temperature of 1 pound of water by 1 °F. 1 BTU = 1055.6 joules

Prehistoric Sources

We live in a universe brimming with energy of all types, and over the millennia humans have enjoyed the benefits of, and have learned to use, many different forms of energy. Before humans knew anything more than basic survival skills, we enjoyed the blessings of sunlight and the food provided by growing plants.

Natural fire caused by lightning strikes was a fact of early existence for several million years, with archaeological evidence indicating that humans deliberately made fire for at least 800,000 years. Fire was used for light and heat, for cooking food, and for protection from dangerous animals. We can easily imagine important social interactions with families or small tribal groups gathered around a communal fire. Later, fire would be used for clearing land for agricultural use, for making ceramics, and eventually for refining metals, fashioning metal tools, and developing other increasingly complex technologies.

About 6000 B.C.E., the first boats were built, using the energy of flowing water to provide transportation. These boats were simple canoes, dugouts, or rafts. The wheel was known since about 4000

B.C.E., although the first wheels were used for making pottery. These potter's wheels were spinning tables on which clay could be molded into bowls, pots, jars, and other containers. The earliest known wheeled vehicles appear about 3500 B.C.E. Various domesticated animals were harnessed to these vehicles, providing the energy source for motion.

Passive solar heating of houses was used in ancient Greece, and later by the Romans, starting about 500 B.C.E. The Greeks used building materials such as sandstone, tile, concrete, and adobe clay plaster that can absorb, store, and release heat. Houses were built with south-facing windows and doors, allowing sunlight to fall on these materials during daylight hours. At night, the released heat helped to keep the house warm.

The ancient Chinese mined coal as long ago as 6000 B.C.E. Coal was difficult to mine, so demand wasn't particularly high until about 1000 C.E., when wood became scarce in many parts of China. Starting about 200 B.C.E., the Chinese used coal for heating and for refining metals.

Sunlight and fire remain important energy sources, but both suffer from the same limitation. Neither, by itself, can produce **mechanical work**. For our purposes, mechanical work occurs when a physical object is moved from one place to another. If I lift a book bag off the floor, walk to my next class, climb a set of stairs, or perform any other kind of physical activity, I am doing mechanical work.

mechanical work: the amount of energy transferred by a force, typically causing an object to move

Early Historic Sources of Mechanical Work

Certainly, the first source of mechanical work would be the individual person, or groups of people. Eventually, domestication of animals allowed their use for mechanical work. Dogs were probably domesticated sometime between 12,000 and 30,000 B.C.E. Draught animals such as oxen (~4000 B.C.E.), donkeys and horses (~3000 B.C.E.), and water buffalo (~2500 B.C.E.) followed. Later, camels (~3000 to 1500 B.C.E.) and elephants (~2000 B.C.E.) were domesticated. These dates are approximate, and the animals selected are representative but not exhaustive.

Animals are limited as a source of work because they can learn only a limited number of tasks, and many tasks are beyond any animal's ability. Humans are much more flexible. Forced labor, slavery, has been a fact of human society for millennia. Slavery predates recorded history, but the Code of Hammurabi, written circa 1755 B.C.E., indicates that slavery had been practiced for thousands of years before the code was written.

In ancient times, people became enslaved for many reasons, such as debt, birth into a slave family, war, or punishment for a crime. Slaves not only performed difficult manual labor but in many cultures held more advanced positions, such as physicians, accountants, engineers, and architects. The kinds of work performed depended on the education level of the slave. In many parts of the world, slavery is still practiced today.

Wind has been an important energy source for millennia. The earliest known sailing ships were used by Egyptians about 4000 B.C.E., and by 200 B.C.E. wind-powered water pumps and windmills for grinding grain were developed. Flowing water was also an important energy source. The first reference to water wheels dates to 500 B.C.E. in China, and to about 300 B.C.E. in Greece. During Roman times, water wheels were used for irrigating crops, grinding grain, and pumping water out of mines, among other applications. When available, wind or water is free and relatively easy to use. However, wind is highly variable, and in many locations flowing water is rare or of insufficient quantity or velocity. A small, slow-flowing stream isn't going to provide sufficient energy to perform work.

Ancient peoples discovered important physical principles and invented machines to help them perform work. Levers, pulleys, inclined planes, wheels and axles, wedges, screws, and more complicated devices incorporating one or more simple machines allowed people to magnify force, or to change the direction of an applied force. However, these machines still required some external energy source to supply the forces being magnified. This situation almost changed in 62 C.E. with the invention of the aeolipile by Hero (or Heron) of Alexandria.

Hero (~10–70 C.E.) was a mathematician and inventor and is often considered one of the greatest experimenters of the ancient world. Among his inventions were the first vending machine, which dispensed holy water when a coin was inserted; an organ operated by a windmill; automatic machinery for the theater; a piston water pump used to fight fires; and automatic temple doors that opened and closed by themselves, and many other devices. All these inventions relied on Hero's detailed, practical knowledge of the forces that can be exerted by air and water.

The aeolipile was a steam-powered engine (Figure 1.1). Water in the bottom bowl was heated by a fire, and the steam flowed through two tubes into a hollow sphere and then out of the two nozzles, causing the sphere to spin rapidly. Had someone found a clever way of using this rotation for work, the Industrial Revolution might have started 1,700 years earlier. However, Hero's invention was considered a toy and a curiosity, and not a practical device. We can only wonder what our world would be like today if Hero's engine had been developed into something useful.

Along the coastline, the ebb and flow of tides have been used as an energy source for centuries, and the earliest tidal mill dates to about 619 C.E. These mills typically had storage ponds, filled by water from the rising tides. When the tide receded, the water flowed out of the pond, turning a water wheel. Tidal mills were relatively common throughout the Middle Ages into the Renaissance and were used for grinding grain.

Our ancestors labored under some severe disadvantages compared to us, but nevertheless managed to develop healthy, prosperous, and reasonably sophisticated civilizations. Many of the modern conveniences we take for granted, such as running water, sanitary sewers, good roads, and a reasonably high standard of living, were achieved by people without sophisticated technology.

Figure 1.1   Hero's Aeolipile, the First Steam-powered Engine

Hero, "Hero's Aeolipile: the First Steam Powered Engine," Knight's American Mechanical Dictionary, ed. Woodcroft, 1876.

Yet, over the years, various authors writing for the mainstream, popular press, have invoked the "ancient astronaut" theory to explain the development of human culture and technology. This theory basically says that our ancestors weren't smart enough to have built pyramids or thriving cultures, or to have come up with creative solutions to problems that seem unsolvable to the authors.

Instead, extraterrestrial beings must be responsible for bringing building technology, certain religious beliefs, and other advanced ideas and technologies to early humans. Probably the most famous of these authors is Erich Anton Paul von Däniken (1935–), whose book *Chariots of the Gods* was a bestseller. In this book, von Däniken claims it would have been impossible for ancient Egyptians to build the pyramids due to their inability to move large stone blocks over great distances and to raise these blocks to the required heights.

But the reality is much simpler: Using simple machines known to be available at the time, ancient Egyptians built some very impressive structures. What is true for the ancient Egyptians is also true for the ancient Mayans, Babylonians, and other ancient peoples. Von Däniken makes factual and logical errors about virtually every example cited in *Chariots of the Gods* and in its sequels, and these errors aren't restricted to the ancient Egyptians, but to nearly every civilization he cites.

This "ancient astronaut" theory is still popular among science fiction fans through television programs such as *Stargate SG-1* and *Battlestar Galactica*. But fiction is fiction, even if it includes "science" in its name. Advocates of these theories clearly don't appreciate the ingenuity and creativity of our ancestors. When you don't have a convenient method for performing some extremely difficult task, then you must be ingenious and creative! Despite their relatively unsophisticated technology, our ancestors were just as smart as we are today, and they solved engineering problems that the average person would find challenging. Just because you can't imagine a solution to a problem doesn't mean that your great-great-great-great-great grandfather couldn't solve it!

engine: a machine designed to convert one or more forms of energy into mechanical energy

People, animals, wind, and water remained the sources of mechanical work for thousands of years, from prehistoric times through the European Renaissance. Toward the end of the Renaissance, inventors began to build devices meant to convert energy from fire into mechanical work. The invention of the **engine** marked the beginning of an industrial revolution that continues to this day.

First Modern Engines

Figure 1.2   Branca's Steam Engine

Giovanni Branca, "Branca Steam Engine," Le Machine, 1629.

In the 17th century, inventors were designing and suggesting various types of engines powered by steam or other substances. In 1629, Giovanni Branca (1572–1645) described a steam engine, in which steam pushed against flat vanes like those on a paddle wheel (Figure 1.2). He suggested that the engine could be used to pump water or perform other work. While Branca designed the engine, he never built one.

In 1678 Christiaan Huygens (1629–1695), a Dutch mathematician and inventor, described a gunpowder engine (Figure 1.3). The engine was a cylinder with a tight-fitting piston. Gunpowder was loaded into the cylinder, and when it exploded it forced the air out, allowing the piston to drop, which raised whatever was attached to it. This machine has certain similarities to modern gasoline- or diesel-fueled internal combustion engines. Whether or not such an engine could have been built is debatable, and the idea of gunpowder-powered machines was soon abandoned when steam power was developed.

In 1679, Denis Papin (1647–1712), a French physicist, invented the steam digester, an early form of the modern pressure cooker; a closed container with a tightly fitting lid and a safety valve to prevent explosions. When water is heated inside the cooker, the pressure increases, raising the boiling point. More heat and a corresponding higher temperature are achieved. Papin noticed that the steam pressure tended to raise the lid of his cooker, and from this he suggested that the force of steam could be used to raise pistons. However, he never built an engine based on this idea.

The first commercially successful steam engine was invented by Thomas Newcomen (1664–1729) in 1712 (Figure 1.4). Unlike earlier devices, this was a true engine containing moving parts that transferred power to other machines. Fire produced steam in the boiler, filling the cylinder and pushing the piston upward. Cold water from outside flowed into the cylinder, condensing the steam, allowing the piston to fall. The up-and-down motion of the piston was transferred to the pump. The Newcomen engine was used to pump water out of coal mines. It was not particularly efficient, but the fuel supply was essentially free, so efficiency wasn't an important issue.

In the 300 years since Newcomen's invention, thousands of increasingly sophisticated devices powered by many different energy sources have been built. Our modern society is powered by energies that our ancestors couldn't imagine and is filled with machines that would amaze, mystify, and probably terrify them. Or maybe not. Arthur C. Clarke (1917–2008), an English science fiction writer, science writer, futurist, inventor, undersea explorer, and television series host, said it best: "Any sufficiently advanced technology is indistinguishable from magic."

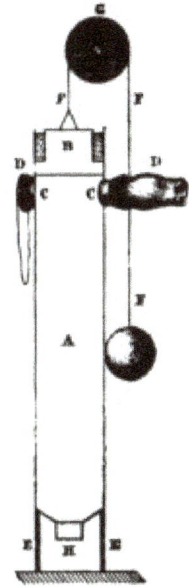

Figure 1.3

Huygens's Gunpowder Engine

Robert Lindsay Galloway, Christiaan Huygens and Willem Jacob Gravesande, "Huygens Gundpowder Engine," The Steam Engine and Its Inventors: A Historical Sketch, 1881.

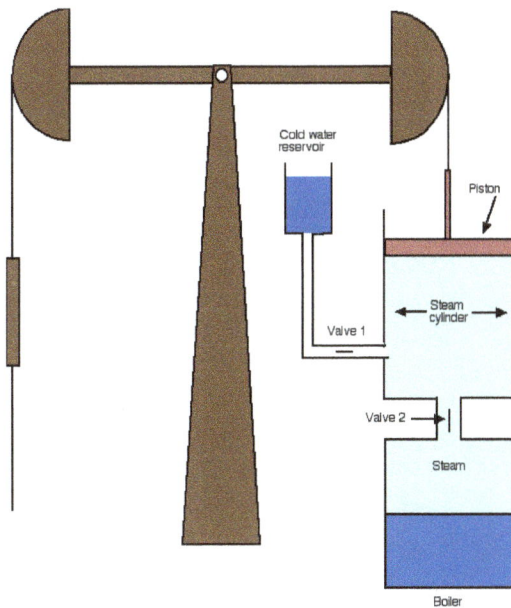

Cold water reservoir

Piston

Steam cylinder

Valve 1

Valve 2

Steam

Boiler

Figure 1.4   Newcomen's Steam Engine

List of Key Takeaways From This Chapter

- Sunlight and fire have been important energy sources for hundreds of thousands of years but could not provide mechanical work.
- Domesticated animals have been an important source of mechanical work but are limited in the kinds of work that can be performed using them.
- Forced labor, slavery, took advantage of the adaptability of humans to perform all kinds of work.
- Wind and water, when available, have been used for mechanical work for thousands of years.
- Engines to convert energy into mechanical work have been known for only the last 300 years.
- Over the last three centuries, humans have developed increasingly sophisticated engines to convert energy into work.

Chapter 1 Exercises

The Internet is a powerful tool for finding answers to many questions, but it must be used intelligently. Not everything on the World Wide Web is factually accurate, and you should consult more than one website. Use the Internet to help you answer the following questions.

1. Imagine you were lost in the woods, without matches or a lighter. What are some ways you might be able to produce fire? What would be required?
2. Other than the animals discussed in this chapter, what other kinds of animals have been used for mechanical work? In what country/culture/society were these animals used?
3. Slavery is still practiced today.
   a. What are the six most common forms of slavery practiced today?
   b. About how many people are currently enslaved?
   c. By number of people enslaved, what are the top five slave-holding countries?
   d. Which international organizations are actively involved in abolishing slavery?

Answers

1. There are three basic ways to make fire without matches or a lighter.
   a. Flint and steel: These are available from outdoor supply or camping stores. If you can find a piece of natural flint, you can then use the blade of a knife for the steel.
   b. Magnifying glass: Just about any glass lens from a camera, glasses, or a real magnifying glass can be used to focus sunlight on a spot.
   c. Friction: This way is the most complicated. You need a curved piece of wood for a bow; a length of cord or string; a long, thin piece of hardwood for a spindle; a hardwood block with a depression/dimple to hold the end of the spindle; and a piece of softwood with a notch cut into it to hold the spindle.
2. In addition to those covered in the chapter, other animals are as follows (country/culture/society included):
   - Yak—Himalayan region, Indian subcontinent
   - Llama—South America
   - Reindeer—Northern Europe, North American (caribou)
   - Zebra—Africa
   - Ostrich—Africa
   - Goats—anywhere that goats can be raised
3. As of 2022,
   a. The most common forms of modern slavery are human trafficking, forced labor, debt or labor bondage, descent-based (children born into slavery), child slavery, and forced or early marriage.
   b. About 50 million people worldwide are currently enslaved (antislavery.org).
   c. These are the top five slave-holding countries: India—18 million; China—3 million; Pakistan—2 million; Bangladesh—1.5 million; Uzbekistan—1.2 million
   d. There are hundreds of organizations. Here are a few:
      - A21 Campaign
      - Better World Campaign
      - Anti-Slavery International
      - Coalition Against Trafficking in Women
      - Free the Slaves
      - Coalition to Abolish Slavery & Trafficking

CHAPTER 2

A Brief Overview of Classical Mechanics

All energy is the same as mechanical energy, whether it exists in the form of motion, or in that of elasticity, or in any other form. The energy in electromagnetic phenomena is mechanical energy.

James Clerk Maxwell

Every one of us is familiar with **mechanical energy**, because every one of us experiences mechanical energy daily. We move our bodies by walking, running, lifting objects, carrying objects. Any kind of activity that involves moving an object from one location to another, by any means whatsoever, involves mechanical energy. Some basic physics ideas are necessary for a clear understanding of mechanical energy.

Learning Objectives

mechanical energy: energy possessed by an object due to its motion or position

This chapter will help students

- Learn the basic ideas and equations governing the motion of objects
- Understand the relationship between work and energy
- Obtain a solid understanding of energy, from a mechanical energy perspective

Metric Units

You have covered the metric system sometime during your elementary and secondary education, so I won't cover this material in detail. I will briefly discuss how the metric system is used in this text.

First, I am not interested in your ability to convert metric units to English units and vice versa. I'm not going to ask you how many inches are in 1 meter, or how many pounds are in 1 kilogram. I am going to use metric units throughout this text.

Second, there are three prefixes commonly used to increase and decrease the size of metric units that you should know:

"kilo": multiply the unit by 1,000
"centi": divide the unit by 100
"milli": divide the unit by 1,000

For example,
1 kilogram (abbreviated kg) = 1 gram × 1,000 = 1,000 grams
1 kilometer (abbreviated km) = 1 meter × 1,000 = 1,000 meters

1 centimeter (abbreviated cm) = 1 meter/100 = 0.01 meter

1 milliliter (abbreviated mL) = 1 liter/1,000 = 0.001 liter

These are generally the only metric prefixes we need in this text. If others are necessary, they will be defined at that time. Additional metric units are discussed when they are introduced.

How Fast?

velocity: the number value of how fast an object is moving, and the object's direction

vector quantity: a quantity that has both magnitude (size) and direction

speed: how fast an object is moving, regardless of the object's direction

We start with *velocity (v)*. Velocity is an example of a *vector quantity*. The formula for calculating velocity is

$$v = \frac{d}{t}$$

where *d* is the distance traveled and *t* is the time. While vector quantities have a direction component, we generally don't specify the direction unless it is necessary. Velocity is often confused with *speed (s)*, which is similar but generally doesn't involve a specific direction. This distinction can be important, but we won't worry about it for now.

Any convenient units of distance and time can be used to determine velocity. In metric units, velocity has units of **meters/second** (m/s).

Knowing any two of these three values allows us to calculate the third value. If I throw a ball a distance of 17 meters, and it takes the ball 2.5 seconds to cover that distance, then the velocity is

$$v = \frac{d}{t}$$

$$v = \frac{17 \text{ m}}{2.5 \text{ s}} = 6.8 \text{ m/s}$$

If I am riding a bicycle at a constant velocity of 35 m/s for 1 hour, how far have I traveled? Well, 1 hour = 60 minutes, and 1 minute = 60 seconds, so in 1 hour I have this:

$$1 \text{ hour } \times \frac{60 \text{ minutes}}{1 \text{ hour}} \times \frac{60 \text{ seconds}}{1 \text{ minute}} = 3,600 \text{ seconds}$$

Since *v = d/t*, I can rearrange the equation:

$$v \times t = d$$

$$35 \tfrac{\text{m}}{\text{s}} \times 3,600 \text{ s} = 126,000 \text{ m}$$

meter: standard metric unit for measuring distance

second: standard metric unit for measuring time

acceleration: the rate of change in velocity occurring in a fixed amount of time

I can always write 126,000 m as 126 km. Either way is fine.

Constant velocity is fine, but we also have changes in velocity. To describe changes in velocity, we need the *acceleration*. Acceleration can be calculated two ways:

$$a = \frac{v}{t} \text{ or } a = \frac{v_f - v_i}{t}$$

In the first equation, Δv means "change in velocity"; in the second equation, $v_f − v_i$ means "final velocity minus initial velocity," which is also "change in velocity." Commonly, Δ (the Greek symbol "delta") means "final value − initial value."

The unit for acceleration is the unit of velocity (m/s), divided by time (s), or m/s^2 (read as "meters per second squared"). While this looks funny, it tells us that in every second, the velocity is changing by some value of meters per second.

Imagine I am sitting in my car at a stoplight. Since the car isn't moving, its velocity is 0 m/s. When the light turns green, I put my foot on the gas pedal and the car starts to move; its velocity increases from 0 m/s. Let's say that after 5 seconds, my car is now moving at 32 m/s. My car's acceleration was

$$a = \frac{32\frac{m}{s}}{5\text{ s}} = \frac{32\frac{m}{s} - 0\frac{m}{s}}{5\text{ s}} = 6.4\ \frac{m}{s^2}$$

In 1 second, my car's velocity went from 0 m/s to 6.4 m/s. In 2 seconds, the velocity increased to 12.8 m/s, in 3 seconds the velocity increased to 19.2 m/s, and after 5 seconds the velocity was 32 m/s.

Acceleration can be positive or negative, depending on whether the final velocity is greater than or less than the initial velocity. Imagine my car is traveling at 55 m/s and I apply the brakes to stop the car at a red light. The final velocity will be 0 m/s. If it takes 6 seconds to stop the car, acceleration is

$$a = \frac{55\frac{m}{s}}{6\text{ s}} = \frac{0\frac{m}{s} - 55\frac{m}{s}}{6\text{ s}} = -9.17\ \frac{m}{s^2}$$

Here the acceleration has a negative sign, indicating that the car's velocity is decreasing. Thus, we use a negative sign to show an opposite effect: If increasing velocity is positive, then decreasing velocity is negative. Like velocity, acceleration is a vector quantity and has direction.

Forces

In general mechanical terms, a *force* causes an object with mass to change its velocity—that is, to accelerate. The size of a force is calculated from the formula

$$F = m \times a$$

Any metric unit of mass can be used, but the standard unit is the kilogram (kg). Using kilogram and m/s^2, we get a force unit of kg-m/s^2 (read as "kilogram meter per second squared"). Please note that the dash between kg and m is not a subtraction sign; it is a hyphen separating two units and is used for clarity.

force: a push or a pull on an object, usually resulting from the object's interaction with another object

newton (N): metric unit of force equal to 1 kg-m/s^2

Alternatively, we could express forces using the *newton (N)*, a metric unit named in honor of Sir Isaac Newton, an English physicist and mathematician.

Forces are vector quantities, and they can be added together if they are in the same direction or subtracted if they are in opposite directions. Figure 2.1 shows two forces being added. The total force acting on the object (the "net" force) is the sum of the two individual forces.

Example A

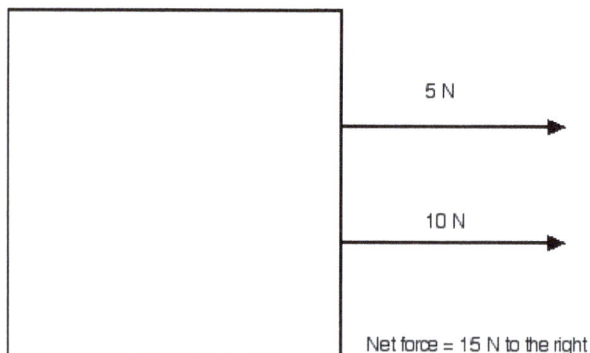

Figure 2.1   Two Forces in the Same Direction Add Directly

Figure 2.2 shows two opposing forces being subtracted. The net force acting on the object is the difference between the two opposing forces.

Example B

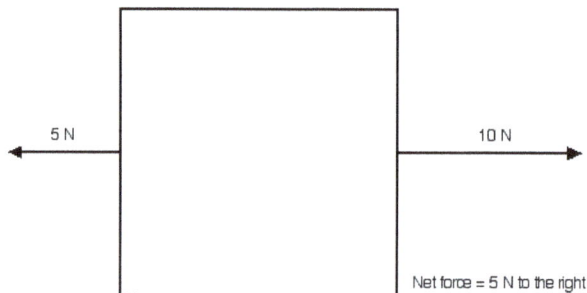

Figure 2.2   Two Opposing Forces Being Subtracted

These are the simplest cases of combining two forces. More complicated combinations require more complicated math, and we don't really need to worry about them for our purposes.

Friction has interesting effects on the motion of objects. While friction tends to reduce the velocity of an object, in many cases friction is absolutely necessary for motion. If it weren't for the friction between our feet and the ground, we wouldn't be able to walk! Anyone who has tried walking on an icy sidewalk knows the importance of friction. Generally, we simplify our calculations by leaving out frictional forces.

friction: a reactive force that resists the motion of an object

When the force is due to gravity, or when we are exerting force against gravity, then the acceleration value used is the gravitational acceleration (abbreviated a_g or just g). While g can change with altitude, we will use the standard value of 9.8 m/s^2. By convention, if the direction is against gravity (the object is moving up), we use -9.8 m/s^2, because gravity causes velocity to decrease. If the direction is with gravity (the object is falling down), we use $+9.8$ m/s^2.

Work and Energy

Mechanical work occurs when a force causes an object to move a given distance, and it is calculated by

$$W = F \times d$$

mechanical work: the amount of energy transferred by a force

joule (J): metric unit of mechanical work

potential energy (E_p): stored energy that depends on the relative position of various parts of the system (mechanical terms)

In the equation, *d* is the distance the object moved and W is the mechanical work performed. Since force (F) has units of kg-m/s^2 or N, and *d* has units of m, work has units of kg-m^2/s^2 (kilogram meter squared per second squared) or N-m (newton meters). We can also use the *joule* (pronounced "jewel"), J, named after James Prescott Joule. The relationship between the joule, the newton, and ordinary metric units is

$$1 \, \text{kg} - \text{m}^2\text{s}^3 = 1 \, \text{N} - \text{m} = 1 \, \text{J}$$

Work and energy are intimately related. A useful definition of energy is "the capacity or ability to perform work." We transfer energy into an object when we perform work on the object. This energy can either be stored in the object or be used to move the object.

For our purposes, the simplest form of *potential energy* (**E$_p$**) is due to an object's position above the ground. Imagine I take a 10-kg object and lift it onto a table that is 1 meter high. How much work have I done?

$$W = F \times d$$

$$W = m \times a_g \times d$$

$$98 \, \text{J} = 10 \, \text{kg} \times 9.8\tfrac{\text{m}}{\text{s}^2} \times 1 \, \text{m}$$

The 98 J of work that I performed on the 10-kg object has been stored as 98 J of potential energy. Notice that the work I performed on the object, against gravity, has been stored in the object as potential energy. Work was converted into potential energy. I can just as easily, and just as correctly, write the formula like this:

$$E_\text{p} = m \times a_g \times d$$

There is also *kinetic energy* (**E$_k$**). Kinetic energy is calculated from the formula

kinetic energy (E_k): energy an object has due to its motion (mechanical terms)

$$E_\text{k} = \frac{1}{2}mv^2$$

Imagine I have a 10-kg object moving at 4.43 m/s. Its kinetic energy would be

$$98 \, \text{J} = \frac{1}{2} \times 10 \, \text{kg} \times \left(4.43\frac{\text{m}}{\text{s}}\right)^2$$

One of the most basic energy ideas is that one form of energy can be converted into another form of energy. If we limit ourselves to mechanical energy, then the two forms are potential energy and kinetic energy.

Law of Conservation of Energy

The law of conservation of energy states that, in a closed system, the total energy at the start is equal to the total energy at the end. Another way of stating the law is that, in a closed system, energy is neither created nor destroyed but can be converted from one form into another. A simple example of this law is illustrated in Figure 2.3.

Let's say my assistant is standing at the top of a building 10 meters tall, holding a 1-kg water balloon. The total mechanical energy of the balloon is the sum of kinetic energy and potential energy. The potential energy is 98 J. The kinetic energy is 0 J, because the balloon isn't moving while my assistant holds it. The law of conservation of energy says that the total energy (98 J + 0 J) at the start is equal to the total energy at the end, and that the total energy doesn't change but can be converted from one form into another.

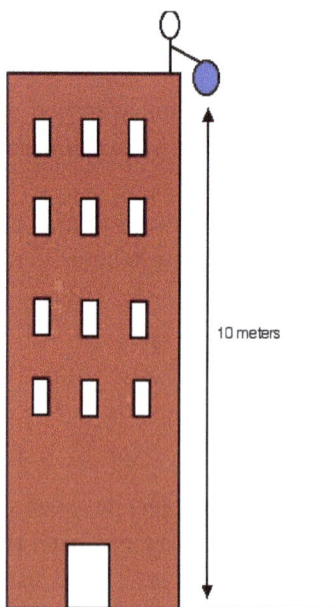

10 meters

Figure 2.3   Water Balloon
Being Dropped Off of a Building

My assistant drops the balloon and it begins to fall. When the balloon is 9 meters above the ground, it has 88.2 J of potential energy, and 9.8 J of kinetic energy, for a total of 98 J.

When the balloon has fallen 5 meters, it is 5 meters above the ground. The potential energy is 49 J and the kinetic energy is 49 J. When the balloon hits the ground, the height of the balloon is 0 meters, so its potential energy is 0 J. Its kinetic energy is 98 J.

Notice that in this example it would be impossible to calculate the kinetic energy from the formula E_k

= ½ mv^2, because we don't know what the velocity of the balloon is at any given height above the ground. We calculate the kinetic energy by knowing (from the law of conservation of energy) that

$$E_T = E_p + E_k$$

As the balloon falls farther and farther, its height above the ground is lower and lower, and therefore its potential energy is smaller and smaller. The total energy (E_T) is constant; therefore the kinetic energy constantly increases.

Power

The last mechanical energy concept is *power*. Power is calculated by

$$P = \frac{E \ (or \ W)}{t}$$

power: time rate of doing work or delivering energy
watt (W): metric unit of power

Units for power can be J/s, or N-m/s, or kg-m^2/s^3, but we commonly use the *watt (W)*. The watt is defined so that

$$1 \ W = 1 \ J/s = 1 \ N - m/s = 1 \ kg - m^2 s^3$$

This is exactly the same unit used to describe all sorts of electrical equipment power usage, including light bulb sizes. A 60-watt lightbulb converts 60 J of electrical energy every second. The unit is named in honor of James Watt, a Scottish inventor best known for his work improving steam engines.

List of Key Takeaways From This Chapter

- Force is required or exerted whenever an object changes its velocity (accelerates).
- Work is performed whenever a force exerted on an object causes the object to move.
- Work and energy are intimately related to each other. In order for work to be performed, energy must be expended. Energy is the capacity (ability) to perform work.
- In a closed system, the total energy remains constant, although the form(s) of energy can change.

Chapter 2 Exercises

Important equations to use and know in answering the following questions:

$$F = ma$$
$$W = Fd$$
$$E_k = \frac{1}{2}mv^2$$
$$E_p = mgh$$
$$P = \frac{W}{t}$$

1. Ignoring friction, how much force is needed to give a 3.0-kg object an acceleration of 5 m/s^2?
2. A force of 2.1 N is exerted on a 7-gram bullet (0.007-kg bullet). What is the bullet's acceleration?
3. A constant force of 1.50 N gives a toy rocket an acceleration of 5.5 m/s^2. What is the mass of the rocket in grams? (1 kilogram = 1,000 grams)
4. A worker pushes horizontally on a large crate with a force of 300 N and the crate moves 5.0 m. How much work was done?
5. A student does 300 J of work in pushing a desk 2.0 meters. How much force did the student exert?
6. A 5.0-kilogram bag of sugar is on a counter. How much work is required to put the bag on a shelf 0.45 m above the counter? (g = 9.8 m/s^2)
7. How much work does gravity do on a 0.150-kg ball falling from a height of 10.0 m (ignore air resistance)?
8. A student throws a 0.150-kg ball straight up, so that it reaches a height of 7.5 m. How much work did the student do?
9. What is the kinetic energy of a 1,000-kg automobile traveling 90 km/h? (There are 1,000 meters in 1 km, and 3,600 seconds in 1 hour.)
10. A 60-kg student riding in a car at constant velocity has kinetic energy of 12,000 J. What is the car's velocity, in km/h?
11. Which has more kinetic energy: a 7.0-gram bullet traveling at 800 m/s or a 4,000-kg ship traveling at 1.0 m/s? Justify your answer.
12. An object is dropped from a height of 12 m. At what height are its kinetic energy and its potential energy the same?
13. A 35-kg child, starting from rest, slides down a waterslide that has a vertical height of 20 m. What is the child's speed when she reaches the bottom of the slide? (Ignore friction.)
14. A 57-kg student climbs a stairway (vertical height of 4.0 m) in 25 seconds. How much work is done? What is the power output of the student?
15. A microwave oven has a power requirement of 1,250 watts. A frozen dinner requires 4.0 minutes to heat on full power. How much energy (joules) is used?

Answers

1. Ignoring friction, how much force is needed to give a 3.0-kg object an acceleration of 5 m/s²?

$$F = ma$$

$$= 3.0kg \times 5^m/_{s^2} = 15N$$

2. A force of 2.1 N is exerted on a 7-gram bullet (0.007-kg bullet). What is the bullet's acceleration?

$$F = ma$$
$$2.1N = 0.007kg \times a$$
$$a = \frac{2.1N}{0.007kg} = 300^m/_{s^2}$$

3. A constant force of 1.50 N gives a toy rocket an acceleration of 5.5 m/s². What is the mass of the rocket in grams? (1 kilogram = 1,000 grams)

$$F = ma$$
$$1.5N = m \times 5.5^m/_{s^2}$$
$$m = \frac{1.5N}{5.5^m/_{s^2}} = 0.273kg = 273g$$

4. A worker pushes horizontally on a large crate with a force of 300 N and the crate moves 5.0 m. How much work was done?

$$W = Fd$$
$$W = 300N \times 5.0m = 1500J$$

5. A student does 300 J of work in pushing a desk 2.0 meters. How much force did the student exert?

$$W = Fd$$
$$F = \frac{W}{d} = \frac{300J}{2.0m} = 150N$$

6. A 5.0-kilogram bag of sugar is on a counter. How much work is required to put the bag on a shelf a distance of 0.45 m above the counter (g = 9.8 m/s²)?

$$W = E_p = mgh$$
$$W = 5.0kg \times 9.8^m/_{s^2} \times 0.45m = 22.05J$$

7. How much work does gravity do on a 0.150-kg ball falling from a height of 10.0 m (ignore air resistance)?

$$W = E_p = mgh$$
$$W = 0.150kg \times 9.8^m/_{s^2} \times 10.0m = 14.7J$$

8. A student throws a 0.150-kg ball straight up, so that it reaches a height of 7.5 m. How much work did the student do?

$$W = E_p = mgh$$
$$W = 0.150kg \times 9.8 \text{ } m/_{s^2} \times 7.5m = 11.025J$$

9. What is the kinetic energy of a 1,000-kg automobile traveling 90 km/h? (There are 1,000 meters in 1 km, and 3,600 seconds in 1 hour.)

$$90km = 90,000m$$
$$1 hour = 3600secs$$
$$90 \text{ } km/_h = \frac{90,000m}{3600s} = 25 \text{ } m/_s$$
$$E_k = \tfrac{1}{2}mv^2 = \tfrac{1}{2}(1000kg)(25 \text{ } m/_s)^2 = 312,500J$$

10. A 60-kg student riding in a car at constant velocity has kinetic energy of 12,000 J. What is the car's velocity, in km/h?

$$E_k = \tfrac{1}{2}mv^2$$
$$12,000J = \tfrac{1}{2}(60kg)v^2$$
$$v^2 = 400 \text{ } m^2/_{s^2}$$
$$v = 20 \text{ } m/_s = 72 \text{ } km/_h$$

11. Which has more kinetic energy: a 7.0-gram bullet traveling at 800 m/s or a 4,000-kg ship traveling at 1.0 m/s? Justify your answer.

$$Bullet: E_k = \tfrac{1}{2}mv^2$$
$$E_k = \tfrac{1}{2}(0.007kg)(800 \text{ } m/_s)^2 = 2240J$$

$$Ship: E_k = \tfrac{1}{2}mv^2$$
$$E_k = \tfrac{1}{2}(4000kg)(1.0 \text{ } m/_s)^2 = 2000J$$

The bullet has more kinetic energy than does the ship.

12. An object is dropped from a height of 12 m. At what height are its kinetic energy and its potential energy the same?

$$E_{total} = E_p + E$$
At the start:
$$E_p = mgh = 1kg \times 9.8 \text{ } m/_{s^2} \times 12m = 117.6J$$
$$E_k = 0J \text{ (the object hasn't dropped yet)}.$$
$$E_{total} = 117.6J + 0J = 117.6J$$
When $E_k = E_p$, $E_p = 58.8J$
$$58.8J = 1.0kg \times 9.8 \text{ } m/_{s^2} \times h$$
$$h = 6.0m$$

13. A 35-kg child, starting from rest, slides down a waterslide that has a vertical height of 20 m. What is the child's speed when she reaches the bottom of the slide? (Ignore friction.)

$$E_{total} = E_p + E$$
At the start:
$$E_p = mgh = 35kg \times 9.8^m/_{s^2} \times 20m = 6860J$$
$$E_k = 0J \text{ (the object hasn't dropped yet)}.$$
$$E_{total} = 6860J + 0J = 6860J$$
When the child reaches the bottom, $E_p = 0J; E_k = 6860J$
$$6860J = \frac{1}{2}(35kg)v^2$$
$$v^2 = 392 : v = 19.8^m/_s$$

14. A 57-kg student climbs a stairway (vertical height of 4.0 m) in 25 seconds. How much work is done? What is the power output of the student?

$$W = E_p = mgh$$
$$W = 57kg \times 9.8^m/_{s^2} \times 4.0m = 2234.4J$$
$$P = \frac{W}{t} = \frac{2234.4J}{25s} = 89.376W$$

15. A microwave oven has a power requirement of 1,250 watts. A frozen dinner requires 4.0 minutes to heat on full power. How much energy (joules) is used?

$$P = \frac{W}{t}$$
$$Pt = W$$
$$1250W \times 240s = 300,000J$$

Hypothesis, Theory, and Law

I have not been able to discover the cause of those properties of gravity from phenomena, and I frame no hypotheses . . .

Sir Isaac Newton

I have taught chemistry in one form or another for more than 40 years. I can immediately distinguish someone who has a reasonably good grasp of science from someone who has a poor grasp of science by asking them to explain some simple, basic terms used in science. The key to understanding science is to think in science terms. The terms *hypothesis*, *theory*, and *law* are probably the most misused and misunderstood words in science, and it is completely unnecessary.

Learning Objectives

This chapter has one learning objective: to understand the meaning of, and to properly use, three terms that are routinely misunderstood and misused in science:

- Hypothesis
- Theory
- Law

Hypothesis

The *Merriam-Webster Dictionary* gives three definitions of **hypothesis**:

hypothesis: a tentative assumption made in order to draw out and test its logical or empirical consequences

1. An assumption or concession made for the sake of argument
2. *A tentative assumption made in order to draw out and test its logical or empirical consequences*
3. The antecedent clause of a conditional statement

The most important of the three definitions is number two. A useful hypothesis is a testable statement, which may involve a prediction. Consider these examples:

1. Too much salt in your diet can contribute to heart disease.
2. The color of light can affect a person's mood.
3. Excessive exposure to sunlight can cause skin cancer.

All these statements make assumptions and provide some sort of test result by which the accuracy of the statement can be judged.

Often, people use *hypothesis* to mean "my (initial) explanation of a phenomenon." This is an archaic use of the term; when Isaac Newton (1643 – 1727), said, "and I frame no hypotheses," he was saying that he couldn't explain what causes gravity. Were he alive today, Newton would probably say, "I have no theory that explains gravity."

In simple terms, a hypothesis assumes or predicts **WHAT** will happen under a given condition but does not explain how or why it happens. If we want some sort of explanation of the phenomenon, then we need something a little more robust than a simple hypothesis.

scientific theory: a detailed and elaborate explanation of a series of apparently disconnected and unrelated observations

Theory

A **scientific theory** is a detailed and elaborate **EXPLANATION** of a series of apparently disconnected and unrelated observations. Typically,

1. A theory is developed after numerous observations are made, by many different scientists, in different laboratories, and after logical and valid reasoning has been applied to these observations.
2. When possible, careful mathematical analysis of experimental results testing the theory is conducted, as part of developing the theory.
3. A useful theory offers predictions that can be tested by other scientists through independent observation and experimentation.
4. A theory must be capable of revision and modification as new observations and experimental results are obtained. A theory should be as generally applicable as possible.
5. A theory must be compatible with other existing theories. If two scientific theories are contradictory, then the contradictions must be resolved either by additional experimentation, or by modification of one or both of the theories, or by combining the two theories into one unified theory, or by rejecting both theories and developing a new theory.
6. Revised and modified theories must explain new observations and experimental results and must be consistent with the observations and results of the original theory.
7. A theory must be falsifiable; that is, as the result of experimentation or observation, it must be possible to show that the theory is wrong.

A scientific theory is a guide to understanding and explaining how and why the natural world works. It is not a simple prediction of an experimental result, it is not a hypothesis that has been "promoted" to a higher level, and it is **NOT AN EDUCATED GUESS!!**

You will frequently hear people who do not know any better dismiss a scientific theory by saying, "Well, that's just a theory." This is the same as dismissing an explanation by saying, "Well, that's just an explanation." What else does an explanation have to be?

Scientific (or Natural) Law

A **scientific (or natural) law** is **NOT** a theory that has been "promoted" to higher status after some period of testing. A scientific law is a principle, a pattern, or a generalization, observed in nature, and normally expressed mathematically. A scientific law allows us to calculate numeric results, and these numerical results allow us to evaluate the scientific law. If it is really a scientific law, then calculated values will match the experimental values.

Many scientific laws have been used to help develop theory, but we can observe a scientific law operating in nature without having any generally accepted theory to explain the law. A classic example of a scientific law without a corresponding theory is Newton's law of gravitation:

> **scientific law:** a principle, pattern, or generalization observed in nature, and normally expressed mathematically

$$F = \frac{Gm_1m_2}{r^2}$$

The exact causes of gravitation were unknown to Newton; he didn't even know the value for G in his equation. Nevertheless, this law has been verified by thousands of observations over the last 400+ years.

Another misconception about scientific law is that it is somehow permanent, unalterable, eternal, and absolutely correct. It is generally true that scientific laws are reliable, given their mathematical nature. However, many scientific laws apply over a limited range of conditions. The gas laws are an example. Gas laws describe the behavior of gases as pressure, temperature, and volume are changed. Gas laws are reliable only until the gas changes into a liquid. Once a gas condenses to a liquid, gas laws no longer apply.

Since laws are based on observations in nature, it is entirely possible that some situations exist that would modify or invalidate a particular scientific law. Humans haven't observed enough of the universe to be absolutely sure of how all parts of the universe function. There may be undiscovered scientific laws that contradict existing scientific laws. I am not saying that this is the case; I am saying you must be open to the possibility. To quote the American theoretical physicist Richard Feynman (1918–1988), "I would rather have questions that can't be answered, than answers that can't be questioned."

It is also possible that some aspects of a scientific law may not be apparent when the law is first announced. A classic example is the case of the law of conservation of energy (already discussed in Chapter 2), and the law of conservation of mass (which you will read about in Chapter 4). Both laws are extremely important in studying chemistry, and both are usually treated separately. However, in 1905, Albert Einstein (1879–1955), also a theoretical physicist, wrote his famous equation:

$$E = mc^2$$

Energy and mass are interchangeable; therefore, the two conservation laws can be expressed as a single law: In a closed system, the sum of energy and mass is constant. Neither law by itself was absolutely correct.

"Proof"

Two words that grate on the nerves of many scientists are *prove* and *proof*, especially when applied to hypothesis, theory, and law.

When an experiment testing a hypothesis is conducted, the result will either support the hypothesis, contradict the hypothesis, or be inconclusive. No experiment, or series of experiments, can ever "prove" a hypothesis. It is always possible that the very next experiment will contradict the hypothesis. It is possible to reject a hypothesis based upon experimental results. It is possible to (tentatively) accept a hypothesis based on experimental results. No hypothesis can ever be proven.

Theories are not "proven" to be true. A theory is used until experiment or observation indicates that the theory doesn't explain the experiment or observation. When that happens, the theory is modified so that all experimental and observational results are explained. Sometimes, a new theory must be developed to replace an outmoded theory. When this happens, the outmoded theory is not "proven wrong"; it is simply replaced by a better theory. Any theory that can't be rejected and replaced by a new theory is not a scientific theory.

Scientific laws are not "proven" to be true. Either the law produces calculated results that match experimental results, or the two results don't match. If the two results don't match, then you don't actually have a scientific law (or you need to conduct your experiments more carefully).

List of Key Takeaways From This Chapter

- No hypothesis is ever "promoted" to the status of theory. Testing hypotheses is one step in the process of developing a theory, but it is not the only step.
- No theory is ever "promoted" to the status of scientific law.
- The simplest definition of *hypothesis* is that it is a testable statement. A hypothesis predicts that "If 'A' happens, then 'B' is the result."
- The simplest definition of *theory* is that it is an explanation of how and why some part of the natural world works.
- The simplest definition of *scientific law* is that it is a generalization, a principle, or a pattern observed in nature. Scientific law allows us to calculate "How much?"
- Hypothesis, theory, and law provide three different kinds of information. They are complimentary, but not interchangeable, and can exist independently of each other.

Chapter 3 Exercises

For each of the following, indicate whether you think the description is a hypothesis or a theory, and justify your answer.

1. Prisoners who learn work skills while in prison will be less likely to commit a crime when they are released.
2. Employees who arrive at work earlier are more productive.
3. The Earth and other planets revolve around the Sun.
4. Tall men make more money than shorter men.
5. Workplace exercise programs and healthier lunch/snack options result in fewer sick days taken and improved mental and physical health among employees.
6. Of students, 70% prefer to use a tablet or laptop computer in class, rather than taking notes with paper and pencil.
7. Freshman students have lower GPAs than sophomore, junior, and senior students.
8. All cats will jump when they see a cucumber.
9. Earth's continents are formed by the movement of tectonic plates over the Earth's surface.
10. All objects with mass are attracted to each other.
11. Acids are substances that produce hydrogen ions when dissolved in water.

Answers

Exercises 1–8 are examples of hypotheses. They are relatively simple, testable statements that predict a particular outcome for a given situation, but they do not explicitly explain how and why the outcome occurs.

Exercises 9 and 11 are examples of theories. In Exercise 9, the reason for the formation of continents is explained as being due to the motion of tectonic plates; this explains why and how continents are formed. In Exercise 11, a specific property (acidity) is associated with a specific situation—the formation of hydrogen ions in water. This explains why one substance is classified as an acid, while another is not.

Exercise 10 should be classified as a hypothesis. It tells what happens, but it doesn't explain how or why it happens (other than being related somehow to having mass).

CHAPTER 4

Fire, Temperature, and Heat

> There is not a law under which any part of this universe is governed which does not come into play and is not touched upon during the time a candle burns.
>
> Michael Faraday

Since its discovery, people have wondered about the nature and causes of fire. Ancient civilizations left writings about fire, but I don't want to discuss these ideas—entire books could be written about them.

Along with fire, the ideas of temperature and heat have a complicated and intertwined relationship, with many common or similar ideas. It is difficult to completely separate these three topics, but teaching them all together can be very confusing. Instead, I'll start with the first "modern" theory about combustion (burning).

Learning Objectives

This chapter will help students

- Understand the combustion process
- Understand how one scientific theory is replaced by a better theory
- Understand the basis of temperature scales
- Understand the modern theory of heat

Two Combustion Theories

Just watching a flame for a few moments leaves us with the impression that something is being released during combustion. For early scientists, this led to the idea of a "fire particle" as part of the fuel, and to the idea that combustion releases this fire particle into the air. "Phlogiston" is the name given to this theoretical fire particle, and the phlogiston theory of combustion was proposed in the late 1600s by Johann Becher (1635–1682) and his student Georg Stahl (1659–1734).

Several important, empirical observations were explained by this theory:

1. During combustion, air is required. The released phlogiston must go somewhere, and the air acts as a receiver for phlogiston.

If a candle is burned in a small, enclosed container, the small volume of air can hold only a small amount of phlogiston. Once the air is filled with phlogiston, combustion stops (the candle burns out).

If a larger container is used, there is more room for phlogiston, so the candle burns longer.

2. Metals heated in air lose their phlogiston, producing a heavier product called a "calx" (similar to rust). From this, it is apparent that phlogiston has negative mass.
3. If calx is heated with a phlogiston-rich material (such as charcoal), the phlogiston flows out of the charcoal and into the calx, and metal is produced. The mass of metal is less than the mass of calx, consistent with phlogiston having negative mass.
4. If a phlogiston-rich substance, such as charcoal, is burned in air, the remaining ash weighs less than the original charcoal. When phlogiston flows out of metals, the residue (calx) is heavier. When phlogiston flows out of charcoal, the residue (ash) is lighter. Evidently, sometimes phlogiston has positive mass.

Stahl and others didn't worry too much about this contradiction. What was important for them was the final effects, not the details.

Not every scientist accepted this theory, mostly because of conceptual problems such as negative mass. One of these scientists was Antoine Lavoisier (1743–1794), one of four scientists known as "the father of modern chemistry" [the other three were John Dalton (1766–1844), Robert Boyle (1627–1691), and Jöns Berzelius (1799–1848)]. Starting in 1772, Lavoisier conducted many combustion experiments, using a good-quality weight scale to measure changes in mass.

One representative experiment involved heating tin to produce tin calx. It was known that tin heated in air would completely convert to calx, because the air could hold all the phlogiston released by the tin. Lavoisier's experiment involved heating tin for different lengths of time in a closed metal flask. Table 4.1 shows the kinds of results obtained by Lavoisier, although I use modern metric units instead of the original units.

Table 4.1. Typical Results Obtained by Lavoisier, in Modern Metric Units

	Before heating	After heating
Flask + contents	110.00 g	110.00 g
Flask + contents after opening (whoosh!)		110.26 g
Empty flask	100.00 g	100.00 g
Tin	10.00 g	8.02 g
Calx	0.00 g	2.24 g
Total contents	10.00 g	10.26 g

First, Lavoisier weighed the flask and its contents before and after heating; in our example the mass was 110.0 g. No matter how long the flask was heated (the longest time was 101 days of continuous heating), the total mass remained constant.

According to the phlogiston theory, if phlogiston is being added or subtracted from the flask and contents, the mass should change. Why doesn't the mass decrease or increase?

Second, when Lavoisier opened the flask to retrieve the contents, he heard a "whooshing" sound. This can only be something entering or leaving the flask. He reweighed the open flask and found it was now 0.26 grams heavier. Perhaps phlogiston released from the tin was trapped in the flask and couldn't escape until it was opened.

Why couldn't the phlogiston escape from the closed flask?

Third, Lavoisier weighed the flask and discovered that nothing had happened to it: The flask weighed exactly the same before and after heating. Phlogiston from the fire had no effect on the flask.

Why wasn't the flask changed from all the phlogiston added to it?

When he compared the mass of tin and calx before and after heating, he discovered that 1.98 grams of tin was converted into 2.24 grams of calx, and the total weight of the contents was 10.26 grams after heating. Before heating, the total contents weighed 10.00 grams.

This odd behavior of phlogiston, sometimes affecting one component but not affecting the others, was very confusing.

Fortunately, during the same time period (from about 1648 to 1785), other scientists were studying gases, their physical and chemical properties, and how to produce them in very pure form. Scientists were also studying the composition of the atmosphere. These studies provided an important insight, leading Lavoisier to propose an alternative explanation for combustion. Instead of the tin losing phlogiston to form the heavier calx, perhaps some part of the air inside the flask combined with the tin to make the calx. What part of the air might this be?

In 1774, Lavoisier received a letter from a Swedish chemist named Carl Scheele (1742–1786) describing a new gas. About a month after receiving the letter, he was visited by an English chemist named Joseph Priestly (1733–1804), who described the same gas and his methods for producing it. Priestly produced this gas by heating mercury oxide (the calx of mercury), producing mercury metal and the gas (Figure 4.1). Lavoisier eventually named this gas "oxygen."

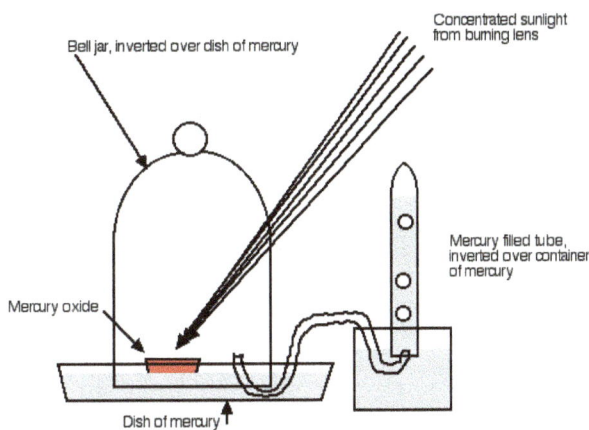

Figure 4.1   Priestly's Method for Obtaining Oxygen from Mercury Oxide

Lavoisier's experiments with mercury oxide cast doubt on the validity of the phlogiston theory. He heated mercury in a closed container. The heated mercury formed mercury oxide (mercury calx). The amount of air in the closed container decreased by a fixed amount, which Lavoisier carefully measured. The amount of air removed was proportional to the amount of mercury calx formed.

Lavoisier took the mercury oxide formed by this experiment and heated it in a closed container, converting the calx back into mercury and gas. He recovered the same amount of gas that was lost during the original heating.

This experiment is strong evidence supporting the idea that oxygen from the air is combining with the mercury, and that the oxygen can be completely released from the mercury oxide formed during the heating. Phlogiston is not needed to explain the experimental results.

Lavoisier used oxygen as the basis for his combustion theory, which is that combustion involves the combination of oxygen with other metals to form calx. When he burned carbon (C), sulfur (S), or phosphorous (P) with oxygen, he produced similar products: carbon dioxide (CO_2), sulfur dioxide (SO_2), and diphosphorous pentoxide (P_2O_5). When these products are dissolved in water, they form acidic solutions. This was why Lavoisier chose the name *oxygen*, from the Greek words "oxys," meaning sharp or sour, and "-gens," meaning born or generated.

The total mass of starting materials (oxygen and other substance) is equal to the total mass of products. The law of conservation of mass is obeyed. There are none of the confusing increases and decreases in mass that are caused by "phlogiston" with its negative or sometimes positive mass.

However, he had a problem with waxes and oils. When candle wax or oil is burned with oxygen, the masses don't add up. The total mass of starting materials is larger than the total mass of products. Lavoisier had no explanation for this discrepancy, and it cast doubt on the validity of the oxygen theory of combustion.

While Lavoisier was performing combustion experiments, an English scientist named Henry Cavendish (1731–1810) was experimenting with a different gas (eventually named *hydrogen*) that was produced by reacting metals with acids. Cavendish burned hydrogen with oxygen, producing water. This is a critical insight that was overlooked by Lavoisier, because of how gas experiments were performed.

In general, when a gas was generated by a chemical reaction, it would be collected by bubbling the gas into a water-filled bottle (Figure 4.2). When Lavoisier conducted his experiments, he used this method to collect any gas products from combustion. However, this method won't work if water is one of the products: It is impossible to distinguish water produced by combustion from water in the bottle. This problem was fixed by using mercury in place of water. Any water produced as a gas would bubble through the mercury.

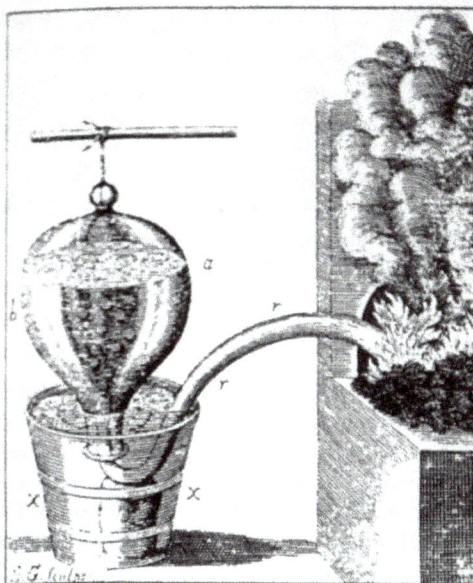

Figure 4.2   Gas Collection by Bubbling into a
Water-filled Bottle

Stephen Hales, https://commons.wikimedia.org/wiki/
File:Stephen_Hale_-_pneumatic_trough.jpg, 1727.

When Lavoisier heard of Cavendish's experiments with hydrogen gas, he understood his oversight: He hadn't accounted for water produced by burning candle wax, oil, and other similar substances. Once the water was included, the law of conservation of mass was obeyed.

Lavoisier's work did not *disprove* the existence of phlogiston, nor did it *disprove* the phlogiston theory of combustion. Lavoisier's oxygen theory of combustion *replaced* the phlogiston theory. The oxygen theory explained all the observations of the phlogiston theory, and obeyed the law of conservation of mass, and allowed for a new understanding of the chemical products produced by combustion. This is the way science works: A better theory replaces an outmoded or flawed theory.

Thermometers and Temperature Scales

For most people in past centuries, temperature was not particularly important, outside of the weather and their personal comfort. Some professions, such as blacksmithing, routinely dealt with high temperatures during the forging of metals. These craftsmen learned through experience how to judge the temperature of hot metal by its color. Mostly, the idea that temperature indicates degrees of "hotness" or "coldness" was sufficient for ordinary uses.

This definition is still the most common, and it is perfectly fine for ordinary social interactions. However, it is not useful in science, because it is subjective: it depends too much on the individual response to a particular temperature. It is also inconsistent. Cold air at 55 °F doesn't provide the same level of "coldness" as cold water at 55 °F.

thermometer: a device used to measure and indicate temperature

freezing point: the temperature at which a given substance changes from liquid to solid

boiling point: the temperature at which a given substance changes from liquid to gas

During the Renaissance, many people built devices to measure temperature changes. The first person to build a reasonably good **thermometer** that gave accurate values was Gabriel Fahrenheit (1686 – 1736). He built his own thermometers and established a temperature scale based on three points. The lowest temperature he could achieve conveniently, using a mixture of ice, water, and salt, was set to a value of 0 (zero). The **freezing point** of rainwater was set at 4, and human body temperature was set at 12.

Why did he pick these values? Why not? One set of values is as good as another.

Eventually, Fahrenheit built better thermometers than his original ones, and he adjusted his temperature scale by multiplying the original values by 8. This assigns the freezing point of water at 32 °F and human body temperature at 96 °F. Eventually, the Fahrenheit scale set the freezing point and **boiling point** of water at 32 °F and 212 °F, respectively.

The next temperature scale was invented by Anders Celcius (1701–1744). Celcius hoped that his temperature scale would become the standard, universally used scale. Originally, his scale set the freezing point of water at 100 °C and the boiling point at 0 °C. Following his death, these values were reversed, making measurements more practical.

Converting between the two temperature scales is not difficult. The formula is

$$°\text{C} = \frac{°\text{F} - 32}{1.8}$$

Water boils at 212 °F. Subtracting 32 from 212 gives us 180, and dividing 212 by 1.8 gives us 100 °C.

$$°100 = \frac{°212 - 32}{1.8}$$

There is a significant problem with either of these scales. Many physical properties depend directly on temperature. One example is the volume of a gas. As the temperature increases, the volume of a gas increases. As the temperature decreases, the volume of the gas decreases. We can describe this relationship, and calculate volume changes, using the formula

$$V = kT$$

V and T are the volume and temperature, and k is the constant that relates these values. Imagine I have 1.00 liter of gas at 25 °C. The value for k is 1.00 liter/25 °C = 0.04 L/°C.

If my temperature increases to 50 °C, then my volume is 2.00 liters (0.04 × 50 = 2.00). If my temperature decreases to 10 °C, then my volume is 0.40 liters (0.04 × 10 = 0.40 liters). If my temperature decreases to 0 °C, then my volume is 0.00 liters. This is reasonable, because I can have a volume of empty space without any matter in it. This is called "vacuum."

However, 0 °C is not the lowest possible temperature; we can have negative temperatures, and this causes the problem. If the temperature drops to –10 °C, then the formula tells us that the volume is –0.40 liters. A negative value for volume has no physical meaning. It is one thing to say my volume decreased by 0.40 liters, it is another to say that I have –0.40 liters. One makes sense, the other doesn't. Since this problem occurs because of how the temperature scales were defined, we can fix the problem by defining a new temperature scale.

This was done in 1848 by William Thomson (1824–1907), also known as Lord Kelvin. Thomson defined a new, absolute temperature scale starting with a value of 0 (zero). Since the scale starts at zero, by definition there cannot be any negative temperatures. When we use the Kelvin scale, we don't get any physically meaningless negative values due to an accident of arithmetic.

Kelvin and Celcius temperatures are related by the formula

$$K = circ\text{C} + 273.15$$

With the lowest possible temperature, by definition, as 0 K, the corresponding Celsius temperature is −273.15 °C, and the corresponding Fahrenheit temperature is −459.67 °F.

Eventually, a nonsubjective, physically meaningful definition of temperature was developed by James Maxwell (1831–1879) and Ludwig Boltzmann (1844–1906) with their development of the Kinetic Molecular Theory of Gases. In this theory, the average kinetic energy of **ideal gas** particles is related to their absolute temperature by the equation

ideal gas: a theoretical gas composed of randomly moving particles that don't interact with each other

$$\overline{E_k} = \frac{1}{2}mv_{rms}^2 = \frac{3}{2}k_B T$$

You've already seen the first part of the equation, calculating the kinetic energy, in Chapter 2. In this equation, we are using the average kinetic energy of all particles of the ideal gas, which we indicate by putting a bar above E_k. The "rms" attached to the velocity term is basically the average velocity of a collection of particles. The second part of the equation contains a constant, k_B, and the absolute temperature, T.

With this formula, we now have a way of relating one form of energy—kinetic energy—to another form of energy represented by temperature. This is much more useful scientifically than describing relative amounts of "hotness" or "coldness."

Modern Heat Theory

As with temperature, most people thought of heat as it applied to their own personal situation. The ideas of temperature and heat are often intertwined: We have all heard people say things like "The temperature is very high, you can really feel the heat." This combination of ideas is perfectly reasonable for ordinary social interactions. However, as with temperature, in science a much more rigorous definition of heat is necessary.

While many scientists thought about heat and discussed it, Joseph Black (1728–1799) was the first to provide many critical insights into heat. He thought of heat as a unique kind of fluid he named "calor." Calor was able to penetrate all material bodies, increasing their temperature. Black conducted several experiments investigating calor.

In one experiment, he mixed one gallon of boiling water with one gallon of ice-cold water and obtained a temperature halfway between boiling and freezing. Clearly, the calor was equally distributed throughout the two gallons of water. He defined a unit of heat as the amount necessary to raise the temperature of 1 pound of water by 1 °F (the calor is the amount of heat needed to raise the temperature of 1 gram of water by 1 °C).

heat capacity: the ratio of heat absorbed by a substance to the temperature change of the substance

When he mixed different materials, he obtained different temperature results. Mixing 1 pound of hot water with 1 pound of cold mercury produced a temperature closer to the hot water than the cold mercury.

From this result, he concluded that different materials required different amounts of heat to change their temperature—the idea of **heat capacity**.

latent heat: the additional heat energy needed to cause a phase change at constant temperature

Another important idea introduced by Black was **latent heat**. When ice at 0 °C melts, it produces liquid water at 0 °C—no temperature change occurs. Similarly, when water at 100 °C boils, it produces steam at 100 °C—again, no temperature change occurs. The additional heat needed to melt ice or boil water is the latent heat. Black envisaged the latent heat loosening the structure of ice, making it a liquid, and loosening the structure of liquid water, making it a gas.

This idea of heat as a fluid was used and extended by other scientists of the time, notably by French engineer Sadi Carnot (1796–1832), often described as the "father of thermodynamics." Carnot compared the operation of a steam engine, where heat energy is converted into mechanical energy, to the operation of a water wheel.

In the water wheel (Figure 4.3), water flows over the wheel, causing it to turn. The amount of work produced in the water wheel depends on the difference in height of the water levels. In the steam engine, heat flows through the engine, performing work. The amount of work produced in the steam engine depends on the difference in temperature between the boiler and the condenser.

Figure 4.3   Water Wheel

Firkin, https://openclipart.org/detail/274921/water-wheel, 2017.

However, Carnot made one mistake in his analysis. In the water wheel, no water is lost as it passes over the wheel. Similarly, Carnot believed the amount of heat entering the condenser was exactly the same as the amount produced by the boiler. The mechanical work was performed as the heat "fell" from high temperature to low temperature. We now know that only part of the heat is transformed into mechanical work and the remainder passes into the condenser (Figure 4.4).

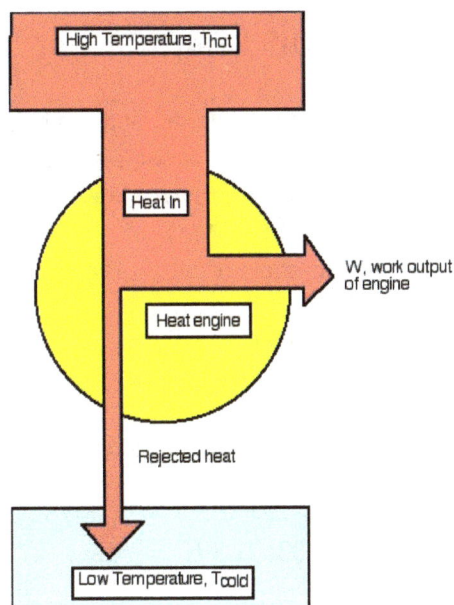

Figure 4.4   Heat Engine (Modified Carnot Description)

A competing heat theory was that heat is caused by some kind of internal motion, and not by a fluid. Probably the first significant advocate of this idea was Benjamin Thompson (1753–1814), later known as Count Rumford. Thompson knew from observation that heat can be produced from nothing—by friction—with no chemical transformation. While watching cannons being bored at an arsenal, he found that the heat produced by friction was seemingly inexhaustible, especially if the boring tool was blunt.

By comparing the heat capacity of the metal block with an equal weight of fragments, Thompson confirmed that no changes had occurred to the cannon material. The heat capacity was identical. He tried to compare the weight of a hot object to a cold object—trying to detect the presence or absence of calor—but didn't get particularly good results. He argued that the apparently indefinite generation of heat by friction was incompatible with the caloric theory, and that the only explanation was that heat was some form of motion.

Other scientists, notably Julius von Mayer (1814–1878) and James Joule (1818–1889), extended Thompson's ideas. Both were interested in the mechanical motion equivalent of heat. Joule's experiments were better known, and von Mayer's contribution is generally overlooked, but both men are now considered pioneers in the development of **thermodynamics**. We will talk more about thermodynamics in the next chapter.

thermodynamics: the branch of science dealing with the relationship between heat and other forms of energy

List of Key Takeaways From This Chapter

- Experimental results can challenge existing theories, and can help produce new scientific theories.
- An obsolete or incomplete theory is not "proven wrong"; it is replaced by a better theory.
- Generally used or generally understood ideas are not always useful in scientific work. Temperature as "hotness" or "coldness" is of limited value.
- If one measurement scale produces unusable results, it is acceptable to define a new measurement scale.
- Heat and temperature are related to each other.
- One kind of energy can be related to a different kind of energy.

Chapter 4 Exercises

$$°C = (°F -- 32)/1.8 \quad K = °C + 273.15$$

1. Perform the indicated temperature conversions:

°F	°C	K
215		
	37	
		950
	212	
77		
		198

2. Arrange the following temperatures from lowest to highest: 25 °C, 74 °F, 325 K.

Answers

1. The missing temperature conversions are shown in colored type:

°F	°C	K
215	101.7	374.8
98.6	37	310.15
1,250.3	676.9	950
413.6	212	485.2
77	25	298.2
−103.3	−75.2	198

2. You can't simply arrange the temperatures by increasing value, because they aren't the same scale. Is 10 inches longer than 3 yards? Of course not. To compare the temperatures, we have to convert them to the same temperature scale.

$$(74\,°\text{F} - - 32)/1.8 = 23.3\,°\text{C}$$
$$325\,\text{K} - - 273.15 = 51.85\,°\text{C}$$

So, 74 °F is the lowest of the three temperatures. 25 °C is the next highest, and 325 K is the highest of the three temperatures.

Calorimetry, Heat Transfer, and Thermodynamics

In all cases in which work is produced by the agency of heat, a quantity of heat is consumed which is proportional to the work done; and conversely, by the expenditure of an equal quantity of work an equal quantity of heat is produced.

Rudolf Clausius

Energy from chemical reactions can be released in several forms: as mechanical energy, as light, as electricity, and as heat. The two most common forms are heat and mechanical energy, and this chapter explores the relationships between calorimetry, the mechanical equivalent of heat, heat transfer, and thermodynamics.

Learning Objectives

This chapter will help students

- Understand the relationship between temperature and heat
- Understand the relationship between mechanical energy and heat
- Understand two fundamental methods of heat transfer
- Understand the laws of thermodynamics and their effects on usable energy

Calorimetry

In developing their heat theories, Black, Thompson, and others performed many experiments using **calorimetry**. Black can be considered the "inventor" of calorimetry, defining many of the ideas we discussed earlier. To perform these experiments, we need some type of **calorimeter**. A simple calorimeter can be made from a polystyrene foam cup and a thermometer.

The basic idea of calorimetry is to allow heat to flow into a fixed mass of water. By measuring the initial and final temperatures of the water, we can calculate the amount of heat absorbed. The important equation is

calorimetry: the act of measuring certain variables for the purpose of determining the heat transferred during chemical reactions, phase changes, or other processes

calorimeter: a device used in calorimetry experiments to measure heat changes during certain processes

specific heat: the amount of heat required to raise the temperature of 1 gram of a material by $1°C$

$$q = mC_{\mathrm{p}}\Delta T$$

where q is the amount of heat, m is the mass, C_{p} is the **specific heat**, and ΔT is the change in temperature (final temperature minus initial temperature). For water, C_{p} = 4.18 J/g-$\overset{\circ}{}$C.

This equation gives us a valuable insight about heat and temperature: They are **NOT** the same thing! Two different amounts of the same material, at the same temperature, have different amounts of heat. Let's look at a simple example: How much heat is needed to raise the temperature of 113 grams of water (about 4 ounces) from 25 $\overset{\circ}{}$C to 75 $\overset{\circ}{}$C?

The ΔT = 75 $\overset{\circ}{}$C – 25 $\overset{\circ}{}$C = 50 $\overset{\circ}{}$C. Using the above equation, we get

$$q = 100 \text{ g } \times\ 4.18 \tfrac{\text{J}}{\text{g} \, ^{\circ}\text{C}} \ \times\ 50 \, ^{\circ}\text{C} = \ 20,900 \text{ J}$$

Let's compare this amount of heat energy to a bigger system: a bathtub containing 30 gallons of water. We'll keep the initial and final temperatures the same—the only change is the mass. One gallon of water is 3,785 grams, so 30 gallons is 113,550 grams. The heat required is

$$q = 113,550 \text{ g } \times\ 4.18 \tfrac{\text{J}}{\text{g} \, ^{\circ}\text{C}} \ \times\ 50 \, ^{\circ}\text{C} = \ 23,731,950 \text{ J}$$

Why the difference? Because we have a larger mass. There are several ways of understanding heat, but one way I've found useful is the definition of temperature from the Kinetic Molecular Theory:

$$\overline{E_{\mathrm{k}}} \ = \ \tfrac{3}{2} k_{\mathrm{B}} T$$

The fraction 3/2 is constant and can be combined with the k_{B} constant into a single value I'll call k. This gives us

$$\overline{E_{\mathrm{k}}} \ = \ kT$$

The average kinetic energy of particles is directly proportional to the temperature—or to put it another way, temperature indicates the average kinetic energy of particles. How do we get an average? Here is a very simple example: Imagine I have 5 particles with the kinetic energies shown below:

Particle number	Kinetic energy
1	17 J
2	8 J
3	39 J
4	115 J
5	1 J

The total kinetic energy of the five particles is 180 J. The average kinetic energy is the total kinetic energy divided by the number of particles, or 180 J/5 = 36 J.

We can think of heat as the "total kinetic energy" of all the particles, while temperature is the average kinetic energy of the particles. This isn't 100% correct, but it does allow us to understand the relationship between heat and temperature, and why different amounts of the same substance require different amounts of heat.

Calorimetry allows us to conduct many interesting heat experiments. One of these is heating ice until it converts into steam. By measuring the heat added and the temperature achieved, we can produce a heating curve (Figure 5.1).

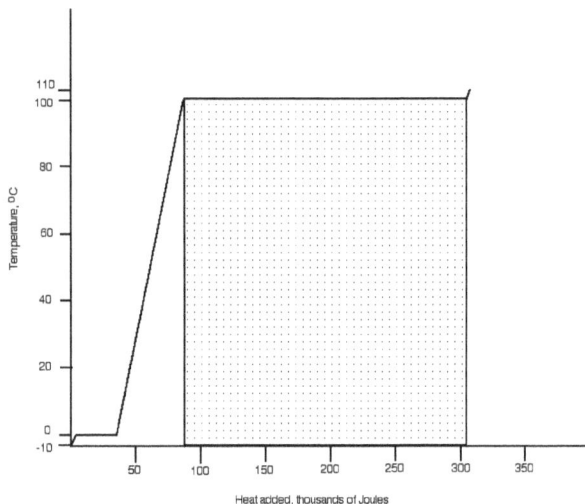

Figure 5.1   Heating Curve for Water

Let's say we have 100 grams of ice at −10 °C. We want to heat it until it converts into steam at 110 °C. How much heat is required?

First, the ice is warmed from −10 °C to 0 °C. The C_p for ice is 2.1 J/g-°C, and the heat required is

$$100g \times 2.1\frac{J}{g\cdot°C} \times 10°C = 2,100J$$

Second, the ice is melted into liquid water. Since there is no temperature change, we must use a different calculation: q = m × L_f, where L_f is the latent heat of fusion. The heat required is

$$100g \times 334\frac{J}{g} = 33,400J$$

The temperature of the water produced from melting the ice is 0 °C. This was an important observation made by Black: Ice melts into water at a constant temperature.

Third, the water is heated from 0 °C to 100 °C. The C_p for liquid water is 4.18 J/g-°C, and the heat required is

$$100g \times 4.18\frac{J}{g\cdot°c} \times 100°C = 41,800J$$

Fourth, the water is boiled, converting it into steam. Again, there is no temperature change, so we use a different calculation. The heat required is found using q = m × L_v, where L_v is the latent heat of vaporization. The heat required is

$$100g \times 2261\frac{J}{g} = 226,100J$$

Finally, the steam is heated to 110 °C. The C_p of steam is 2.1 J/g-°C. By coincidence, this is the same value as the specific heat of ice. The heat required is

$$100g \times 2.1\frac{J}{g \cdot °C} \times 10°C = 2,100J$$

Let's compare the various amounts of heat needed at each stage of our experiment:

Process	Heat required
Warm ice	2,100 J
Melt ice into water	33,400 J
Heat water from 0 °C to 100 °C	41,800 J
Boil water into steam	226,100 J
Heat steam to 110 °C	2,100 J

The single largest energy requirement is converting water at 100 °C to steam at 100 °C. Steam contains lots of energy—over five times as much energy as the same mass of water. This is what makes steam such a useful medium for converting heat energy into mechanical energy—steam contains more energy, so more energy is available for work.

Mechanical Equivalent of Heat

Independent of each other, at about the same time (1842–1843), Joule and von Mayer both proposed the idea that heat and mechanical work are equivalent. Joule's work is more widely known than von Mayer's, and Joule's experiments were very important in establishing the kinetic theory over the caloric theory. Joule performed many different experiments relating mechanical energy to heat energy. One of his best known is shown in Figure 5.2.

In this experiment, a falling weight caused a set of paddles to stir water in an insulated container. Friction between the paddles and the water caused the water temperature to increase. Joule measured the amount of heat and calculated the amount of mechanical energy from the falling weight. The only source of heat energy was the mechanical energy, so he could directly relate the two. The modern standard value is 4.186 J of mechanical work = 1 calorie of heat. Of course, it is much simpler to just use Joule's measurement for energy, regardless of form.

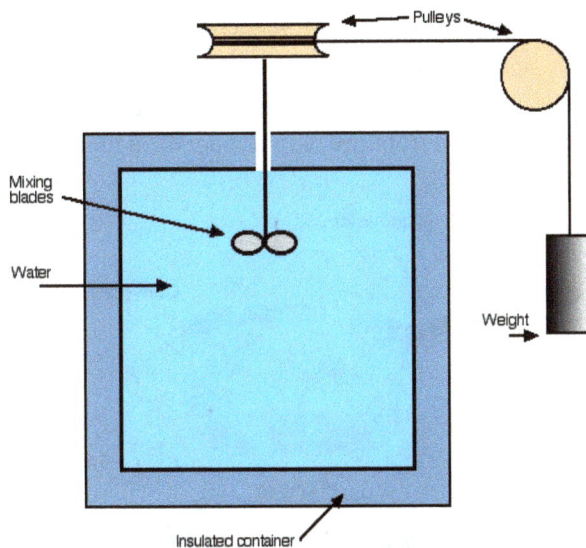

Figure 5.2   Joule's Experiment for the Mechanical
Equivalent of Heat

Heat Transfer

For thousands of years, people have noticed that heat (however they chose to describe it) spontaneously moved from hot to cold, from high temperature to low temperature. We have all had hot food grow cold while it was sitting on the table. The heat from the food flowed into the plate, the table, the surrounding air, the environment. Similarly, we have all had cold food become warm while it was sitting on the table; ice cream melts, cold beverages become warmer. In both cases, heat flowed from higher temperature to lower temperature.

We might not be able to understand this observation if temperature is defined as "hotness" or "coldness," but with temperature being related to average kinetic energy, we can easily understand heat flow. High kinetic energy particles (high temperature) collide with low kinetic energy particles (low temperature). During this collision, energy is transferred from high energy to low energy. Eventually, given enough time and enough collisions, the energy of all the particles will be more evenly distributed. When there is no average kinetic energy difference between the particles (no temperature difference), heat flow stops.

There are two basic heat transfer mechanisms that operate by collisions between particles. The first is **conduction**. Anyone who has touched a hot frying pan, a hot pizza pan, any hot object, already knows about conduction. The hot, high kinetic energy particles collide with cooler, low kinetic energy particles, and energy is transferred by the collision. The handle of a cast iron frying pan becomes very hot, even though the handle is not directly over the burner, because of conduction.

The second mechanism is **convection**. This mechanism (Figure 5.3) is a little more complicated than conduction. Convection involves three steps:

conduction: the transfer of thermal energy by the collision of microscopic particles. The particles can be atoms, molecules, or electrons.

convection: the transfer of thermal energy due to the movement of a fluid

1. Heat is transferred by conduction from a high-temperature source to a low-temperature fluid. The fluid can be either a liquid or a gas.
2. The heated fluid physically moves from one location to another location.
3. Heat is transferred by conduction from the high-temperature fluid to a low-temperature object.

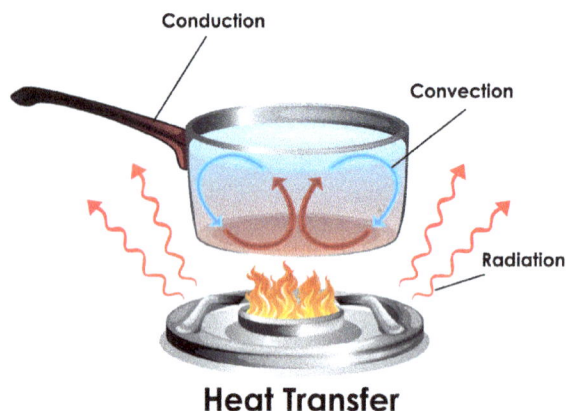

Heat Transfer

Figure 5.3   Convection

Copyright © 2019 Depositphotos/blueringmedia.

In either case, collision between high kinetic energy particles and low kinetic energy particles transfers the heat energy from high temperature to low temperature.

Thermodynamics

Thermodynamics is the branch of physics dealing with the relationship between all forms of energy. Carnot, Maxwell, Boltzmann, and many others made important contributions to thermodynamics and thermodynamic theory. One extremely important individual was Rudolf Clausius (1822–1888). Clausius was responsible for the first explicit statement of the first law of thermodynamics, and he first stated the basic ideas of the second law of thermodynamics.

Here are two reasonable versions of the first law of thermodynamics:

1. The total energy of an isolated system is constant, although energy can be transformed from one form to another.
2. Energy can neither be created nor destroyed (in an isolated system).

entropy: a physical property most commonly associated with randomness or disorder

Whichever version you prefer, the first law tells us that the maximum amount of energy for work is limited to the maximum amount of energy available. If we have 100 J of energy, we can do 100 J of work, but not the smallest bit more than 100 J.

The second law of thermodynamics is a little more difficult to describe. It establishes the idea of **entropy**. The simplest statement of the second law is this: "In an isolated system entropy will not spontaneously decrease; it can only increase or at best remain constant." As a consequence of the second law, it is impossible to completely convert 100 J of energy into 100 J of work: There will always be some energy wasted as entropy. The only way for entropy to decrease is for energy to be transferred into the system.

The third law of thermodynamics was developed by Walther Nernst (1864–1941) from 1906 to 1912. In 1912, Nernst stated the law as "It is impossible for any procedure to lead to the isotherm T = 0 in a finite number of steps." The temperature referred to is 0 K, or absolute zero—by definition, the lowest temperature possible. Entropy is affected by the temperature: The greater the temperature, the greater the entropy. If the temperature were absolute zero, then entropy would also be zero. However, since it is impossible to reach T = 0 in a finite number of steps, it is impossible to ever make entropy "go away," and you are stuck with the first and second laws.

There is also the zeroth law of thermodynamics, which states, "If two thermodynamic systems are in equilibrium with a third system, then they are in equilibrium with each other." Generally, this is a mathematical equivalence of the form A = B, B = C, therefore A = C.

If all these scientific laws only confuse you, perhaps "Ginsberg's Theorem" will help you. Named for poet Allen Ginsberg (1926–1997), who is best known for the poem "Howl," the theorem restates the consequences of the four laws of thermodynamics on the usable energy of a system:

1. There is a game (a consequence of the zeroth law).
2. You can't win the game (the consequence of the first law).
3. You can't break even (the consequence of the second law).
4. You can't even get out of the game (the consequence of the third law).

For a poet, Ginsberg seems to have known his thermodynamics pretty well.

List of Key Takeaways From This Chapter

- Calorimetry experiments show us the relationship between heat and temperature.
- Heat and temperature are related, but they are not the same. Heat depends on the temperature and the amount (mass) of substance present.
- Different masses of the same material at the same temperature will contain different amounts of heat.
- A large mass of material at a low temperature can contain more heat than a small mass of the same material at a higher temperature.
- Mechanical energy and heat energy have a fixed relationship to each other, and they can be interchanged with each other.
- Two common methods of heat transfer involve collisions between particles, with the transfer of kinetic energy between the colliding particles.
- Thermodynamics is governed by four basic laws that describe how energy can be converted among its various forms.

Chapter 5 Exercises

$$q = mC_p \Delta T \qquad q = m \times L_f \qquad q = m \times L_v$$

C_p of water is 4.18 J/g-°C.

1. How much heat is needed to melt 25.00 grams of ice at 0 °C, making water at 0 °C. The latent heat of fusion for ice is 334 J/g.
2. A coffee cup contains 118 grams of water. A bathtub contains 159,000 grams of water. If the starting temperature in each case is 25 °C, what is the final temperature of each after 25,000 joules of energy has been added?
   a. Coffee cup final temperature
   b. Bathtub final temperature
3. The specific heat of copper is 0.385 J/g-°C. A mass of copper at 37 °C was dropped into 1,000 grams of water at 25 °C. The temperature of the water rose to 27 °C. What was the mass of the heated copper?

Answers

1. How much heat is needed to melt 25.00 grams of ice at 0 °C making water at 0 °C. The latent heat of fusion for ice is 334 J/g.

$$q = m \times L_f$$
$$= 25.00 \text{ g} \times 334 \text{ J/g}$$
$$= 8{,}350 \text{ joules}$$

2. What is the final temperature of each after 25,000 joules of energy has been added?
   a. Coffee cup final temperature:

$$q = mC_p\Delta T$$
$$25{,}000 \text{ joules} = 118 \text{ g} \times 4.18 \text{ J/g-°C} \times \Delta T$$
$$\Delta T = 50.7 \text{ °C}$$
$$T_f = 25 \text{ °C} + 50.7 \text{ °C}$$
$$= 75.7 \text{ °C}$$

   b. Bathtub final temperature:

$$q = mC_p\Delta T$$
$$25{,}000 \text{ joules} = 159{,}000 \text{ g} \times 4.18 \text{ J/g-°C} \times \Delta T$$
$$\Delta T = 0.038 \text{ °C}$$
$$T_f = 25 \text{ °C} + 0.038 \text{ °C}$$
$$= 25.038 \text{ °C}$$

3. The heat absorbed by the water was

$$q = 1{,}000 \text{ g} \times 4.18 \text{ J/g-°C} \times 2 \text{ °C}$$
$$= 8{,}360 \text{ J}$$

This energy was given up by the heated copper:

$$8{,}360 \text{ J} = m \times 0.385 \text{ J/g-°C} \times 10 \text{ °C}$$
$$m = 8{,}360 \text{ J}/(0.385 \text{ J/g-°C} \times 10 \text{ °C})$$
$$= 2{,}171 \text{ g}$$

CHAPTER 6

A General Overview of Chemistry

> Chemistry is the study of matter. But I prefer to see it as the study of change.
>
> Walter White

This chapter provides a condensed general overview of chemistry. In a regular chemistry course, much more material would be covered over a 2-semester sequence. Our coverage won't be as extensive as you'd get in a regular, general chemistry course, but it should be enough for you to understand the rest of the book.

Learning Objectives

This chapter will help students

- Understand the most common model of the atom
- Distinguish between atoms and molecules
- Understand the difference between ionic compounds and covalent compounds
- Be able to classify matter
- Understand the information provided by a chemical equation
- Balance a chemical equation
- Determine the standard amount of a substance

Atomic Theory

For hundreds of years, philosophers debated the existence of a finite, physical, ultimate particle making up all matter. The Greek philosopher Democritus (~460–370 B.C.E.) proposed an atomic theory very similar to the modern theory. According to Democritus, atoms are individual, indivisible, indestructible particles with empty space separating them. Atoms are in constant, eternal motion; there are an infinite number and infinite kinds of atoms; and atoms differ in shape and size.

Other philosophers opposed Democritus's theory, and debates for and against the theory occupied the philosophers' time over many pleasant afternoons. There was absolutely no way to experimentally test the theory, so there was no basis to reject or confirm it.

For hundreds of years, atomic theories came and went, influenced heavily by alchemy and alchemical ideas. I won't cover alchemy. While it is the direct ancestor of modern chemistry, it's like learning that your great-grandfather was a horse thief: You don't really want to talk about it.

In the early 1800s, John Dalton (see Chapter 4) revived and refined Democritus's atomic theory. Dalton used atomic theory to explain results from gas experiments, to understand the atmosphere and weather (Dalton was a proficient meteorologist), and to calculate relative atomic weights of the various elements. There are four basic components of this theory:

1. Elements are composed of individual small particles (atoms).
2. All atoms of the same element are identical. The atoms of one element are different from the atoms of any other elements.
3. Atoms are indestructible and cannot be created.
4. Compounds are formed when atoms of different elements combine in simple ratios.

Components 1 and 4 remain true today; 2 and 3 have undergone some slight modifications.

chemical properties: a characteristic of a substance that can be observed in a chemical reaction
physical properties: a property of matter not involving a chemical change or chemical reaction
chemical formula: a description of the kinds of elements and the number of atoms of each element that are present in a molecule of a compound

Different elements can have different, or similar, **chemical properties**. Carbon and sulfur react with oxygen, while argon and helium won't react with oxygen. Different elements can have different, or similar, **physical properties**. Hydrogen, oxygen, and nitrogen are gases at room temperature, while iron is a solid at room temperature and mercury is a liquid. Finding at least one chemical or physical property that uniquely identifies each element would be extremely useful.

Dalton (and others) settled on relative atomic weight as the property. The atomic weight of the lightest element, hydrogen, was set to a value of 1.00. When hydrogen combines with oxygen to form water, 1.00 gram of hydrogen requires exactly 8.00 grams of oxygen, therefore the relative atomic weight of oxygen is 8.00. If hydrogen and carbon react to form methane, 1.00 gram of hydrogen combines with 4.00 grams of carbon, and therefore the atomic weight of carbon is 4.00.

Determining relative atomic weight was very difficult, and sometimes it took decades before reliable values were obtained. Part of the difficulty was determining the **chemical formula** of a substance, which was also extremely difficult. For example, Dalton assumed water contained one hydrogen atom and one oxygen atom. It took other experiments to demonstrate that water contains two hydrogen atoms and one oxygen atom. This meant that the atomic weight of oxygen must be 16.00 instead of 8.00.

Dalton's image of an atom was a hard, round, indestructible ball, with a fixed, unique weight. This simple atomic model lasted for about 100 years. Eventually, more facts about atoms were discovered, leading to new atomic models. Probably the most important and most widely taught model is the Bohr model, developed by Niels Bohr (1885–1962) (Figure 6.1).

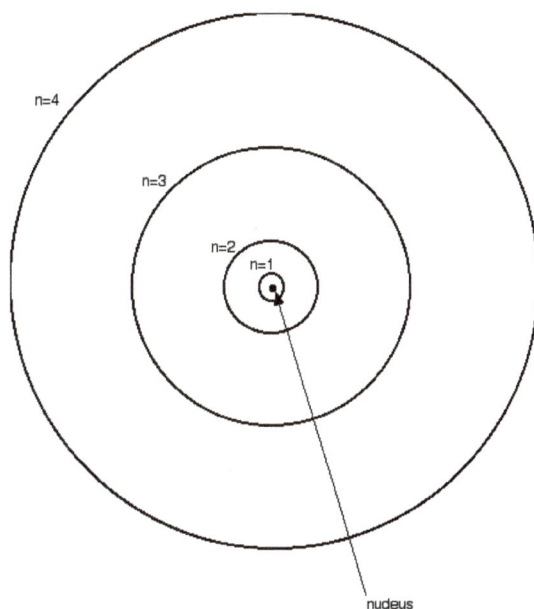

Figure 6.1   Bohr Model of the Atom

The Bohr model shows three basic components of the atom. **Protons** and **neutrons** are grouped together in a **nucleus**. **Electrons** orbit around the nucleus at fixed distances, much like the planets orbiting the Sun. The Bohr model is sometimes called the planetary model because of its general resemblance to the solar system.

Elements

I've talked about elements for several pages and it's time I defined the term. An *element* is one of 118 pure substances that can't be chemically converted into two or more simpler substances. Elements are the primary components of matter, and each element has a unique **atomic number**. All atoms with the same atomic number are atoms of the same element; atoms with different atomic numbers are from different elements. The periodic table of the elements (Figure 6.2) arranges the elements based on their atomic number and chemical properties.

proton: a subatomic particle with a positive electric charge
neutron: a subatomic particle with no electric charge
electron: a subatomic particle with a negative electric charge
nucleus: the positively charged central core of an atom, containing protons and neutrons and containing nearly all of the mass
atomic number: the number of protons present in the nucleus of an atom

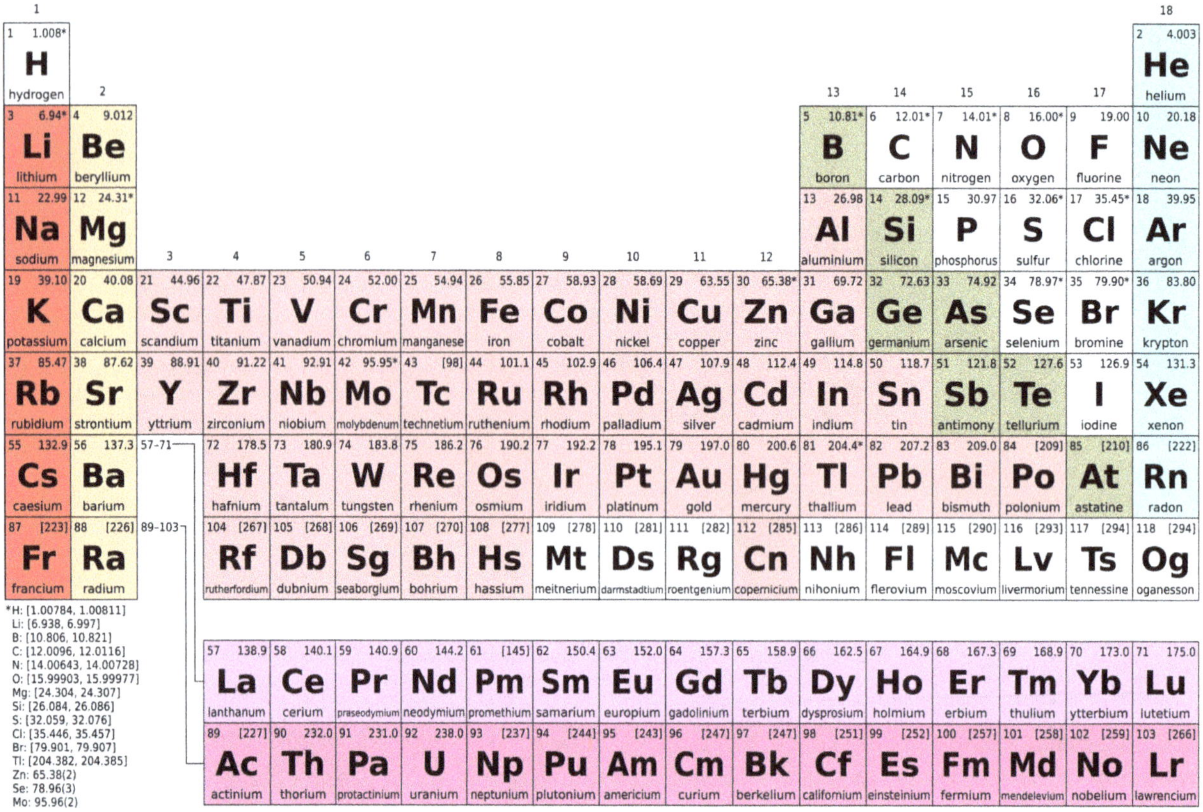

Figure 6.2   The Periodic Table of the Elements

László Németh, "Periodic Table of the Elements," https://commons.wikimedia.org/wiki/File:Periodic_table_simple_en.svg, 2013.

Each element has three unique designators: the atomic number, the element name, and the chemical symbol. Elements are usually named by the person(s) who first discover the element. No two elements have the same name, just as no two elements have the same atomic number. The name and atomic number for an element are fixed and constant: All atoms that have atomic number "1" are atoms of the element "hydrogen."

molecule: two or more atoms chemically combined into a larger particle

chemically combined: two or more substances joined in such a way that they cannot be separated by physical methods

The chemical symbol is a one- or two-letter abbreviation of the chemical name: "H" for hydrogen, "He" for helium, "O" for oxygen, and so on. Some elements have chemical symbols that don't correspond to their current names. Iron is an example: Its chemical symbol is Fe. Many elements have been known for hundreds or thousands of years. Originally, iron was known by its Latin name, *ferrum*, so its symbol was Fe. Several other elements have chemical symbols that come from their original names in Latin or Greek. The symbol for tungsten is W, because the original German name was *wolfram*.

My advice is not to worry about this too much. The important takeaway is that the chemical symbol is also tied to the atomic number and the chemical name. Any one of these three designators can be used to identify a specific element.

Atoms **chemically combine,** forming larger particles called **molecules**. Sometimes all the atoms in a molecule are the same. Two atoms of oxygen combine forming an oxygen molecule, or three atoms of oxygen combine forming an ozone molecule. Ozone and molecular oxygen are different forms of the element oxygen.

If atoms of two or more different elements chemically combine, the result is a molecule of a **compound**. No one knows how many compounds exist, but certainly there are many millions of them. Compounds have a fixed chemical formula, and at least one unique name (a name belonging to only one specific compound). For example, water has the chemical formula H_2O, indicating that a molecule of water contains two atoms of hydrogen and one atom of oxygen, chemically combined.

compound: a pure substance composed of identical molecules formed from two or more elements

Ionic and Covalent Compounds

Two basic mechanisms cause atoms to chemically combine into molecules, and both mechanisms depend on the electrons orbiting the nucleus. Look at Figure 6.1 carefully. The protons and neutrons are locked up in the nucleus of the atom. Several layers of electrons surround the nucleus, preventing the nucleus from interacting with another atom. The electrons are a different matter, especially those electrons in the **valence shell**. Electrons in the valence shell can directly interact with the valence shell of a second atom.

One mechanism involves one or more electrons transferring between two atoms. Normally, an atom has equal numbers of protons and electrons. The total negative charge from the electrons is equal to the total positive charge from the protons, and the atom is electrically neutral. If an atom loses one electron, its total negative charge is reduced by one unit. The positive charge from the protons in the nucleus isn't changed—no proton was lost. Therefore, losing one electron produces a **cation** with a net +1 electric charge.

Some atoms gain one or more electrons, increasing their total negative charge. Gaining one electron produces an **anion** with a net −1 electric charge.

valence shell: in an atom, the outside layer of electrons

cation: a single atom or small molecule having a positive electrical charge

anion: a single atom or small molecule having a negative electrical charge

A simple example of electron transfer can be seen with the elements sodium (Na) and chlorine (Cl). Sodium loses one electron, producing a sodium cation (Na^+):

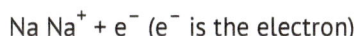

$$Na \quad Na^+ + e^- \quad (e^- \text{ is the electron})$$

Chlorine gains one electron, producing a chloride anion (Cl^-):

$$Cl + e^- \quad Cl^-$$

There is a very, very powerful electrostatic attraction between any two objects having opposite electric charges. The sodium cation and the chloride anion are strongly attracted to each other, forming a molecule with the formula NaCl:

$$Na^+ + Cl^- \quad NaCl$$

ionic bonding: forming a linkage by electrostatic attractions between oppositely charged ions in a chemical compound

ionic compounds: substances containing anions and cations, held together by electrostatic attractions

This mechanism is called **ionic bonding**, and it produces **ionic compounds**.

Initially, all bonding was believed to be ionic bonding, but this led to inconsistent and confusing results. For example,

- In NaCl, Na is "+" and Cl is "−"
- In HCl, Cl is "−"; therefore H must be "+"

- In CH_4, H is "+"; therefore C must be "–" (actually –4)
- In CCl_4, C is "–4" and Cl is "–"; but this is impossible!

When is the element an anion and when is it a cation? Various rules were adopted, but there were numerous exceptions and a great deal of general confusion.

In 1916, Gilbert Lewis (1875–1946) proposed a new bonding theory. Instead of electrons being transferred between two atoms, a pair of electrons could be shared between two atoms, forming a **covalent bond**. Molecules formed by covalent bonds are called "covalent compounds" and are the vast majority of known chemical compounds.

> **covalent bond**: linkage between atoms formed by sharing a pair of electrons

Whether the compound is ionic or covalent, molecules can be described using their chemical formula. Chemical formulas follow a simple pattern: Each element in the molecule is identified using its chemical symbol. The symbol is followed by a subscripted number, indicating how many atoms of the element are present. If there is only one atom of the element, no number is given (we don't use a subscript 1).

Some common examples:

- NaCl—one sodium atom and one chlorine atom
- H_2SO_4—two hydrogen atoms, one sulfur atom, four oxygen atoms
- $C_6H_{12}O_6$—six carbon atoms, twelve hydrogen atoms, six oxygen atoms

Chemical formulas provide useful information, but they are limited to indicating kinds of elements and number of atoms—they can't tell us how the atoms are connected together to make specific compounds. Figure 6.3 shows how two entirely different molecules can have the same chemical formula.

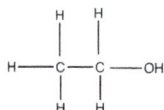

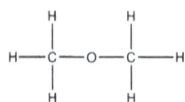

Figure 6.3   Two Different Molecules With the Formula C_2H_6O

Classification of Matter

All matter can be divided into two general categories: pure substances and mixtures.

Mixtures are simple physical combinations of two or more substances. Neither of the substances is chemically altered during mixing, and the substances can be separated by physical processes. Sugar dissolved in water is a mixture, and neither the sugar molecules nor the water molecules are changed during the mixing process. We can recover the sugar by allowing the water to evaporate.

Mixtures can be divided into two categories: homogeneous mixtures and heterogeneous mixtures. "Homogeneous" and "heterogeneous" describe the bulk characteristics of the mixture and do not apply at the atomic level or the molecular level. Whether at the atomic level or the molecular level, **ALL** mixtures are heterogeneous.

A homogeneous mixture has a uniform, constant composition and only one phase of matter. Southern-style sweet tea is a homogeneous mixture. Every sip has the same amount of sugar (sweetness), tea, and water. There is no undissolved solid sugar at the bottom of the glass.

A heterogeneous mixture has nonuniform composition, and typically two phases. Chicken noodle soup is a heterogeneous mixture. One spoonful may have only chicken broth, while another spoonful may have mostly noodles, with very little broth. A liquid phase (the broth) and a solid phase (noodles, chicken, vegetables) are present. The solids tend to be at the bottom of the bowl.

Sometimes, a heterogeneous mixture can be made temporarily homogeneous. Oil and vinegar salad dressing can be shaken very vigorously, mixing the oil and vinegar uniformly. However, if the dressing is then allowed to sit undisturbed, the oil and vinegar separate into two distinct liquid layers. The homogeneous mixture returns to its original heterogeneous form.

Pure substances are divided into two categories: elements and compounds. These have already been discussed. Figure 6.4 summarizes the classification of matter.

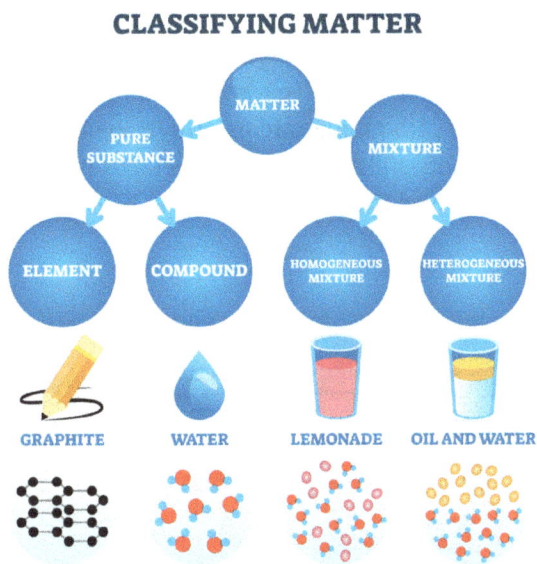

CLASSIFYING MATTER

Figure 6.4   Classification of Matter

Copyright © 2020 Depositphotos/VectorMine.

Chemical Equations

Chemical reactions result in substances changing into new substances. These changes are described using a **chemical equation**. Here is a typical chemical equation:

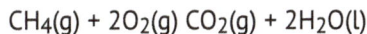

$$CH_4(g) + 2O_2(g) \rightarrow CO_2(g) + 2H_2O(l)$$

To write a chemical equation, we first write all compounds that are **reactants** on the left side, followed by the **reaction arrow**, and then write the **products** on the right side. In our example, CH_4 and O_2 are reactants and CO_2 and H_2O are products. Sometimes additional information is given in parentheses following the chemical formula. In our example, (g) indicates that the substance is a gas, while (l) indicates a liquid.

chemical equation: a representation of a chemical reaction, using chemical formulas to identify the substances, and a set of numbers and special symbols

reactant: a substance that participates, and undergoes a chemical change, in a reaction

products: a substance obtained as the result of a chemical reaction

There are two kinds of numbers shown in the chemical equation. We have subscripts as part of the chemical formula, discussed earlier. We also have **coefficients** in front of the chemical formula, indicating how many molecules of the compound are reacting.

A *balanced* chemical equation is a statement of the law of conservation of mass. To be balanced, the equation must meet two criteria:

1. The same kinds of elements must be present as reactants and products.
2. For each element, the same number of atoms must be present as reactants and products.

Our sample equation is a balanced chemical equation. Carbon (C), hydrogen (H), and oxygen (O) are the only elements involved and are present as both reactants and products. One atom of carbon, four atoms of hydrogen, and four atoms of oxygen are present as reactants and as products. Sometimes it is helpful to draw pictures of the molecules involved in a chemical reaction (Figure 6.5).

reaction arrow: a symbol separating reactants and products in a chemical reaction. It can be interpreted to mean "yields," "produces," "is converted into," or a similar meaning.

coefficient: a number placed in front of a chemical symbol or formula, indicating the number of atoms or molecules participating in a chemical reaction

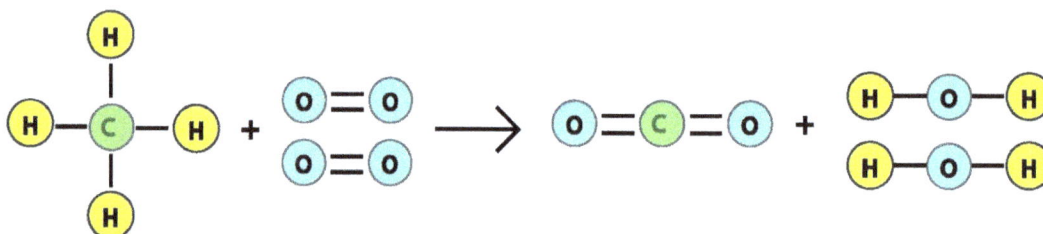

Figure 6.5   Chemical Reaction Using Pictures of Molecules

The process of balancing a chemical equation is not too difficult using a simple set of steps. For example, the steps below describe the process for this equation:

$$C_6H_6 + O_2 \quad CO_2 + H_2O$$

Step 1. Compare the elements in the products and reactants. We have carbon, oxygen, and hydrogen on both sides of the equation, which is critical. If we had another element, such as iron (Fe), on one side but not the other, then there is nothing we could do to balance the equation. We would have to give up.

Step 2. Generally, it is best to start with an element that is present in one reactant and in one product. Carbon or hydrogen is present in one reactant (C_6H_6). Carbon is present in CO_2 (only), while hydrogen is present in H_2O (only). Oxygen would be a poor choice to start with, because it is present in both products.

Step 3. Picking carbon, we compare the number of carbon atoms in C_6H_6 (6 carbons), with the number of carbon atoms in CO_2 (1 carbon). The only way to make the number of carbon atoms identical is to put a coefficient of 6 in front of CO_2.

Note: Changing the subscripts in the chemical formula is **NEVER** an option!

Now the carbon atoms are balanced, and the equation is

$$C_6H_6 + O_2 \quad 6CO_2 + H_2O$$

Step 4. Now we balance the hydrogen. There are 6 hydrogens in C_6H_6, and 2 hydrogens in H_2O. A coefficient of 3 is needed in front of H_2O.

$$C_6H_6 + O_2 \quad 6CO_2 + 3H_2O$$

Step 5. Finally we balance the oxygen. There are 2 oxygen atoms on the left side, and 15 oxygen atoms on the right. We could put a coefficient of 7.5 in front of oxygen:

$$C_6H_6 + 7.5O_2 \quad 6CO_2 + 3H_2O$$

The problem with "7.5" is that it indicates one-half of a molecule, which cannot be. We fix this issue by multiplying the entire reaction by 2:

$$2C_6H_6 + 15O_2 \quad 12CO_2 + 6H_2O$$

As a check, compare the total numbers of atoms of each element on the left and right sides:

Element	Left side	Right side
Carbon	12	12
Hydrogen	12	12
Oxygen	30	30

Standard Amount of a Substance

The standard amount of substance is the **mole**. In simple terms, a mole is a counting number, used for counting groups of objects. Just like we count eggs by the dozen, shoes by the pair, years by the score, and sheets of paper by the ream, we count atoms, ions, molecules, and other subatomic particles by the mole. The mole is a specific number of these particles, called "Avogadro's number": 6.02214×10^{23} objects = 1 mole.

mole: the amount of a substance that contains as many particles as there are atoms in 0.012 kg of carbon-12
formula weight: the sum of the atomic weights of all atoms appearing in a chemical formula

While the mole is a specific number of particles, we don't actually count out individual atoms or molecules. Instead, we use the atomic or formula weight of a substance. When I have a mass, in grams, of a substance equal to the atomic or **formula weight** of the substance, then I have one mole of the substance. For example, the formula weight of H_2O is 18.0148. If I have 18.0148 grams of water, then I have 1 mole of water, and therefore 6.02214×10^{23} molecules of water. Similarly, 58.443 grams of sodium chloride (NaCl) is 1 mole of NaCl, and 180.1548 grams of sugar ($C_6H_{12}O_6$) is 1 mole of sugar.

For any given mass of a substance, we can calculate the number of moles of the substance using the formula

$$\text{number of moles} = \frac{\text{mass in grams}}{\text{formula (or atomic) weight}}$$

How many moles of water are present in 200 grams of water?

$$\text{number of moles of water} = \frac{200 \text{ grams of water}}{18.0148} = 11.102 \text{ moles of water}$$

How many grams of sugar are present in 5.00 moles of sugar?

$$5.00 \text{ moles of sugar} = \frac{\text{mass of sugar}}{180.1548}$$

$$5.00 \text{ moles of sugar} \times 180.1548 = 902.74 \text{ grams of sugar}$$

List of Key Takeaways From This Chapter

- All matter is divided into pure substances and mixtures.
- Pure substances are either elements or compounds.
- An element has identical atoms.
- A compound has identical molecules.
- Molecules are two or more atoms chemically combined.
- Mixtures are either homogeneous or heterogeneous. Homogeneous mixtures have one phase and constant composition; heterogeneous mixtures have two or more phases and variable composition.
- An atom is the smallest particle of an element, and it contain protons, neutrons, and electrons.
- The outside shell of electrons (valence electrons) is responsible for chemical bonding.
- Ionic bonding occurs when one or more valence electrons are transferred between two atoms.
- Covalent bonding occurs when one pair of electrons is shared between two atoms.
- Molecules are described by their chemical formula.
- The chemical formula shows the number of each kind of atom present in a molecule.
- A chemical reaction converts one substance into a different substance.
- Chemical reactions are described using a balanced chemical equation.
- The mole is the fundamental metric unit for the amount of substance.

Chapter 6 Exercises

1. Classify each of the following as either a pure substance or a mixture. (Depending on exact circumstances, some materials could be either.)

Type of matter	Pure substance	Mixture
1. Chlorine gas		
2. Water		
3. Soil		
4. Sugar water		
5. Oxygen gas		
6. Carbon dioxide gas		
7. Rocky road ice cream		
8. Rubbing alcohol		
9. Pure air		
10. Iron		
11. Milk		
12. Gasoline		
13. Caesar salad		
14. Gold		
15. Hydrochloric acid		
16. Bacon		
17. Brewed coffee		
18. White gold		

2. Balance the following chemical equations. Detailed answers are given at the end of this set of exercises. Don't look at the answers until you have completed the problems.
   a. $Fe + O_2 \rightarrow Fe_2O_3$
   b. $C_2H_4 + O_2 \rightarrow CO_2 + H_2O$
   c. $P_2O_5 + H_2O \rightarrow H_3PO_4$
   d. $Al_4C_3 + H_2O \rightarrow Al(OH)_3 + CH_4$
   e. $NH_4NO_3 \rightarrow N_2 + O_2 + H_2O$

f. $Na_2O + H_2O$ NaOH

g. $Na_2SiO_3 + HF$ $H_2SiO_3 + NaF$

h. $C_3H_5N_3O_9$ $CO_2 + N_2 + O_2 + H_2O$

i. $NaHCO_3 + H_3C_6H_5O_7$ $CO_2 + H_2O + Na_3C_6H_5O_7$

j. $KClO_3$ $KCl + O_2$

3. **Formula weight**. Use your periodic table to calculate the formula weight (to two decimal places) of the substances shown below.

$CaCl_2$		NH_4NO_3		$C_6H_{12}O_6$	
$KMnO_4$		UF_6		CH_3OH	
$ZnSO_4$		$NiBr_2$		H_2O_2	
C_6H_5Cl		$AgNO_2$		KI	
C_2H_3O		$Mg_3(PO_4)_2$		Fe_2O_3	

4. **Grams to moles, moles to grams**. For the following compounds, use the given information to calculate either moles or grams as appropriate.

$CaCl_2$	5.00 grams = ___ moles	H_4NO_3	0.20 moles = ___ grams
$KMnO_4$	0.75 moles = ___ grams	UF_6	135.00 grams = ___ moles
$ZnSO_4$	12.35 grams = ___ moles	$NiBr_2$	4.88 moles = ___ grams
C_6H_5Cl	1.50 moles = ___ grams	$AgNO_2$	1255 grams = ___ moles
C_2H_3O	0.73 grams = ___ moles	$Mg_3(PO_4)_2$	0.8855 moles = ___ grams

Answers

1. Classify each of the following as either a pure substance or a mixture. (Depending on exact circumstances, some materials could be either.)

Type of matter	Pure substance	Mixture
1. Chlorine gas	X	
2. Water	X	X
3. Soil		X
4. Sugar water		X
5. Oxygen gas	X	
6. Carbon dioxide gas	X	
7. Rocky road ice cream		X
8. Rubbing alcohol		X
9. Pure air		X
10. Iron	X	
11. Milk		X
12. Gasoline		X
13. Caesar salad		X
14. Gold	X	
15. Hydrochloric acid		X
16. Bacon		X
17. Brewed coffee		X
18. White gold		X

2. Balance the following chemical equations.

    a. $Fe + O_2 \rightarrow Fe_2O_3$

    Balance Fe: $2Fe + O_2 \rightarrow Fe_2O_3$

    Balance O: $2Fe + 1.5O_2 \rightarrow Fe_2O_3$

    Final: $4Fe + 3O_2 \rightarrow 2Fe_2O_3$

    Check: 4 Fe, 6 O vs. 4 Fe, 6 O

b. $C_2H_4 + O_2 \ CO_2 + H_2O$

        Balance C: $C_2H_4 + O_2 \ 2CO_2 + H_2O$

        Balance H: $C_2H_4 + O_2 \ 2CO_2 + 2H_2O$

        Balance O: $C_2H_4 + 3O_2 \ 2CO_2 + 2H_2O$

        Check: 2 C, 4 H, 6 O vs. 2 C, 4 H, 6 O

c. $P_2O_5 + H_2O \ H_3PO_4$

        Balance P: $P_2O_5 + H_2O \ 2H_3PO_4$

        Balance H: $P_2O_5 + 3H_2O \ 2H_3PO_4$

        Balance O: $P_2O_5 + 3H_2O \ 2H_3PO_4$ (nothing changes)

        Check: 2 P, 8 O, 6 H vs. 2 P, 8 O, 6 H

d. $Al_4C_3 + H_2O \ Al(OH)_3 + CH_4$

        Balance Al: $Al_4C_3 + H_2O \ 4Al(OH)_3 + CH_4$

        Balance C: $Al_4C_3 + H_2O \ 4Al(OH)_3 + 3CH_4$

        Balance H: $Al_4C_3 + 12H_2O \ 4Al(OH)_3 + 3CH_4$

        Balance O: $Al_4C_3 + 12H_2O \ 4Al(OH)_3 + 3CH_4$ (nothing changes)

        Check: 4 Al, 3 C, 12 O, 24 H vs. 4 Al, 3 C, 12 O, 24 H

e. $NH_4NO_3 \ N_2 + O_2 + H_2O$

        Balance N: $NH_4NO_3 \ N_2 + O_2 + H_2O$ (nothing changes)

        Balance H: $NH_4NO_3 \ N_2 + O_2 + 2H_2O$

        Balance O: $NH_4NO_3 \ N_2 + 1/2O_2 + 2H_2O$

        Final: $2NH_4NO_3 \ 2N_2 + O_2 + 4H_2O$

        Check: 4 N, 6 O, 8 H vs. 4 N, 6 O, 8 H

f. $Na_2O + H_2O \ NaOH$

        Balance Na: $Na_2O + H_2O \ 2NaOH$

        Balance H: $Na_2O + H_2O \ 2NaOH$ (nothing changes)

        Balance O: $Na_2O + H_2O \ 2NaOH$ (nothing changes)

        Check: 2 Na, 2 O, 2 H vs. 2 Na, 2 O, 2 H

g. $Na_2SiO_3 + HF \ H_2SiO_3 + NaF$

        Notice: SiO_3 is present on both sides. Treat as a single unit.

        Balance SiO_3: $Na_2SiO_3 + HF \ H_2SiO_3 + NaF$ (nothing changes)

        Balance Na: $Na_2SiO_3 + HF \ H_2SiO_3 + 2NaF$

        Balance F: $Na_2SiO_3 + 2HF \ H_2SiO_3 + 2NaF$

        Balance H: $Na_2SiO_3 + 2HF \ H_2SiO_3 + 2NaF$ (nothing changes)

        Check: 2 Na, 1 Si, 3 O, 2 H, 2 F vs. 2 Na, 1 Si, 3 O, 2 H, 2 F

h. $C_3H_5N_3O_9 \ CO_2 + N_2 + O_2 + H_2O$

        Balance C: $C_3H_5N_3O_9 \ 3CO_2 + N_2 + O_2 + H_2O$

        Balance N: $C_3H_5N_3O_9 \ 3CO_2 + 1.5N_2 + O_2 + H_2O$

        Balance H: $C_3H_5N_3O_9 \ 3CO_2 + 1.5N_2 + O_2 + 2.5H_2O$

        Balance O: $C_3H_5N_3O_9 \ 3CO_2 + 1.5N_2 + 0.25O_2 + 2.5H_2O$

        Final: $4C_3H_5N_3O_9 \ 12CO_2 + 6N_2 + O_2 + 10H_2O$

        Check: 12 C, 20 H, 12 N, 36 O vs. 12 C, 20 H, 12 N, 36 O

i. $NaHCO_3 + H_3C_6H_5O_7$ $CO_2 + H_2O + Na_3C_6H_5O_7$

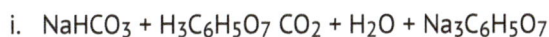

> Notice: $C_6H_3O_7$ is present on both sides. Treat as a single unit.
>
> Balance $C_6H_3O_7$: $NaHCO_3 + H_3C_6H_5O_7$ $CO_2 + H_2O + Na_3C_6H_5O_7$ (nothing changes)
>
> Balance Na: $3NaHCO_3 + H_3C_6H_5O_7$ $CO_2 + H_2O + Na_3C_6H_5O_7$
>
> Balance C: $3NaHCO_3 + H_3C_6H_5O_7$ $3CO_2 + H_2O + Na_3C_6H_5O_7$
>
> Balance H: $3NaHCO_3 + H_3C_6H_5O_7$ $3CO_2 + 3H_2O + Na_3C_6H_5O_7$
>
> Balance O: $3NaHCO_3 + H_3C_6H_5O_7$ $3CO_2 + 3H_2O + Na_3C_6H_5O_7$ (nothing changes)
>
> Check: 3 Na, 11 H, 9 C, 16 O vs. 3 Na, 11 H, 9 C, 16 O

j. $KclO_3$ $Kcl + O_2$

> Balance K: $KclO_3$ $Kcl + O_2$ (nothing changes)
>
> Balance Cl: $KclO_3$ $Kcl + O_2$ (nothing changes)
>
> Balance O: $KclO_3$ $Kcl + 1.5O_2$
>
> Final: $2KclO_3$ $2Kcl + 3O_2$
>
> Check: 2 K, 2 Cl, 6 O vs. 2 K, 2 Cl, 6 O

3. Formula weight. Use your periodic table and calculate the formula weight (to two decimal places) of the substances shown below.

$CaCl_2$	110.98	NH_4NO_3	80.04	$C_6H_{12}O_6$	180.16
$KMnO_4$	158.03	UF_6	352.02	CH_3OH	32.04
$ZnSO_4$	161.45	$NiBr_2$	218.50	H_2O_2	34.01
C_6H_5Cl	112.56	$AgNO_2$	153.87	KI	166.00
C_2H_3O	43.04	$Mg_3(PO_4)_2$	262.86	Fe_2O_3	159.69

4. Grams to moles, moles to grams. For the following compounds, use the given information to calculate either moles or grams as appropriate.

$CaCl_2$	5.00 grams = 0.045 moles	HNO_3	0.20 moles = 12.60 grams
$KMnO_4$	0.75 moles = 118.52 grams	UF_6	135.00 grams = 0.384 moles
$ZnSO_4$	12.35 grams = 0.0765 moles	$NiBr_2$	4.88 moles = 1,066 grams
C_6H_5Cl	1.50 moles = 168.84 grams	$AgNO_2$	1255 grams = 8.16 moles
C_2H_3O	0.73 grams = 0.017 moles	$Mg_3(PO_4)_2$	0.8855 moles = 232.8 grams

CHAPTER 7

Enthalpy, Entropy, and Gibbs Free Energy

> A mathematician may say anything he pleases, but a physicist must be at least partially sane.
>
> J. Willard Gibbs

Chemical energy is contained in chemical bonds attaching atoms together. Making and breaking chemical bonds involves energy changes. If more energy is released in making a new bond than is consumed in breaking an existing bond, then the energy difference is available for use. If more energy is consumed in breaking an existing bond than is released in making a new bond, then energy is absorbed into the resulting molecule.

In this chapter we look at the factors determining the amount of useful energy a chemical reaction can produce.

Learning Objectives

This chapter will help students

- Calculate the enthalpy change in a chemical reaction
- Calculate the entropy change in a chemical reaction
- Calculate the Gibbs free energy of a chemical reaction
- Identify a reaction as exothermic or endothermic
- Identify the reaction as spontaneous or nonspontaneous
- Make meaningful comparisons of energy release by different reactions
- Make meaningful comparisons of energy content of different fuels

standard enthalpy of formation: a measure of the energy released or absorbed when one mole of a substance is created under standard conditions from its pure elements. Units are typically kJ/mole.

Calculating Enthalpy from Standard Values

It is extremely difficult, and unnecessary, to look at the total energy content of a chemical compound. Instead, we determine the energy change during a chemical reaction. To calculate this change, we need to consider two competing natural phenomena: the tendency of systems to have the lowest possible energy, and their tendency to have the greatest possible entropy. Energy and entropy are determined separately, then combined to determine the total energy change during the chemical reaction.

The energy change (actually, the enthalpy of the reaction, ΔH^0) is calculated from the **standard enthalpy of formation** (ΔH_f^0) for each substance in the reaction. Standard enthalpy values are determined either experimentally or by calculation using other standard enthalpy values. For example, we can burn pure carbon with pure oxygen in a calorimeter to form carbon dioxide:

$$C(s, graphite) + O_2(g)\ CO_2(g)$$

Carbon as solid graphite and oxygen as a gas are in their standard elemental forms at 25 °C and 1 atmosphere of pressure. The energy released by this reaction is the ΔH_f^0 for carbon dioxide gas:

$$\Delta H_{rxn} = \Delta H_f^0 \text{ for } CO_2$$

The ΔH_f^0 value for any element in its standard elemental form is zero (0) and is always for one mole of substance, having units of kJ/mole. By convention, if heat is released (temperature increases), then the sign of ΔH_f^0 is negative. If heat is absorbed (temperature decreases), then the sign is positive.

The specific form and phase of substance is important! Carbon as graphite has ΔH_f^0 = 0 kJ/mole, but carbon as diamond has ΔH_f^0 = 1.89 kJ/mole. Diamond is slightly less stable than graphite, therefore its energy is higher. Water vapor, liquid water, and ice have different ΔH_f^0 values, so you must be careful to use the value for the form indicated by the chemical equation.

The enthalpy of reaction (ΔH^0) is the total ΔH_f^0 of the products minus the total ΔH_f^0 of the reactants:

$$\Delta H^0 = \sum \Delta H_f^0 \text{ (products)} - \sum \Delta H_f^0 \text{ (reactants)}$$

Let's look at the combustion of octane:

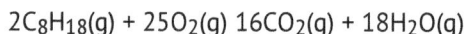

$$2C_8H_{18}(g) + 25O_2(g)\ 16CO_2(g) + 18H_2O(g)$$

These are the ΔH_f^0 values for our substances:

Substance	ΔH_f^0
$C_8H_{18}(g)$	−249.9 kJ/mole
$O_2(g)$	0 kJ/mole
$CO_2(g)$	−393.509 kJ/mole
$H_2O(g)$	−241.818 kJ/mole

We multiply each ΔH_f^0 by the coefficient from the balanced chemical equation:

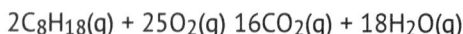

$$2C_8H_{18}(g) + 25O_2(g)\ 16CO_2(g) + 18H_2O(g)$$

(2 × −249.9 kJ/mole) + (25 × 0 kJ/mole) (16 × −393.509 kJ/mole) + (18 × −241.818 kJ/mole)

(−499.8 kJ) + (0 kJ) (−6,296.114 kJ) + (−4,352.724 kJ)

−225.8 kJ −10,648.838 kJ

$$\Delta H^0 = \sum \Delta H_f^0 \text{ (products)} - \sum \Delta H_f^0 \text{ (reactants)}$$

$$\Delta H^0 = (-10,648.838 \text{ kJ}) - (-499.8 \text{ kJ})$$

$$\Delta H^0 = -10,149.038 \text{ kJ}$$

Notice that ΔH^0 has units of kJ, not kJ/mole. Our calculation accounted for all moles of reactants and products. The negative sign of ΔH^0 indicates that energy is released, and the reaction is **exothermic**. If the sign is positive, then energy is absorbed, and the reaction is **endothermic**. Figure 7.1 compares exothermic and endothermic reactions.

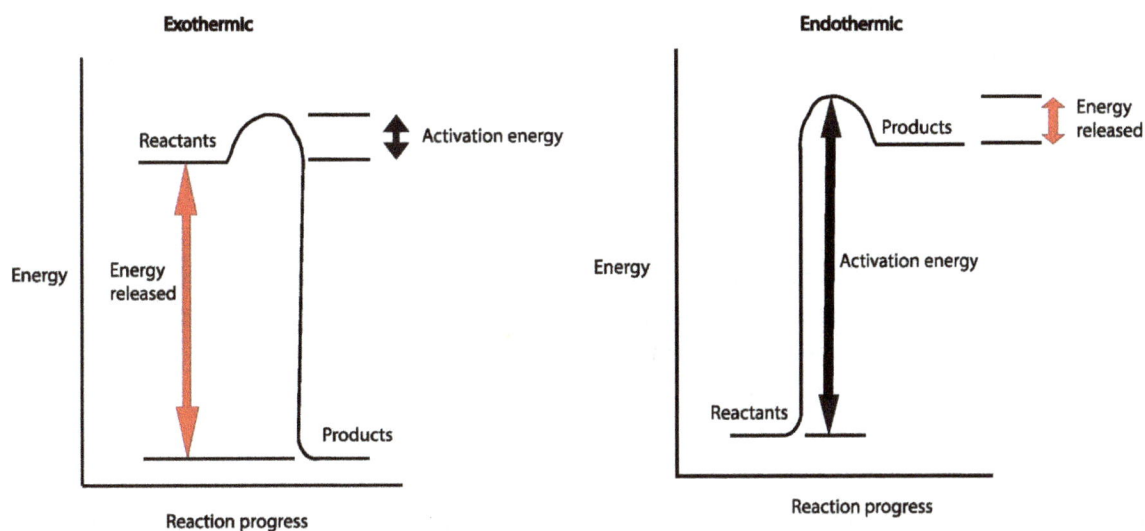

Figure 7.1   Exothermic Reaction versus Endothermic Reaction

If we wrote the chemical equation backward, all our calculations would follow the same procedure. The difference would be the sign of ΔH^0, which would be positive for the reverse reaction.

Calculating Entropy From Standard Values

Entropy calculations follow the same procedure as enthalpy calculations, using **standard molar entropy** (S^0) values to calculate ΔS^0, the change in the entropy of the system.

These are the S^0 values for octane combustion:

exothermic: a reaction that gives off energy

endothermic: a reaction that absorbs energy

standard molar entropy: the randomness of one mole of substance under standard conditions. Units are typically J/moleK.

Substance	S^0
$C_8H_{18}(g)$	361.1 J/moleK
$O_2(g)$	205.138 J/moleK
$CO_2(g)$	213.74 J/moleK
$H_2O(g)$	188.825 J/moleK

Calculating ΔS^0 gives us this:

$$2C_8H_{18}(g) + 25O_2(g)\ 16CO_2(g) + 18H_2O(g)$$

$$(2 \times 361.1 \text{ J/moleK}) + (25 \times 205.138 \text{ J/moleK})\ (16 \times 213.74 \text{ J/moleK}) + (18 \times 188.825 \text{ J/moleK})$$

$$(722.2 \text{ J/K}) + (5{,}128.45 \text{ J/K})\ (3{,}419.84 \text{ J/K}) + (3{,}398.85 \text{ J/K})$$

$$5{,}850.65 \text{ J/K}\ 6{,}818.69 \text{ J/K}$$

$$\Delta S^0 = \sum S^0 \text{(products)} - \sum S^0 \text{(reactants)}$$

$$\Delta S^0 = 6{,}818.69 \text{ J/K} - 5{,}850.65 \text{ J/K}$$

$$\Delta S^0 = 968.04 \text{ J/K}$$

Notice that ΔS^0 has units of J/K; our calculations accounted for all moles of reactants and products. The sign of ΔS^0 is positive, indicating increased entropy. A negative sign for ΔS^0 indicates decreased entropy. Temperature affects the size of the entropy change.

Gibbs Free Energy—Putting It All Together

The total energy available to perform work is the Gibbs free energy (ΔG), named for Josiah Gibbs (1839–1903), a pioneer in chemical thermodynamics. The crucial difference between ΔG and ΔH^0 is that ΔH^0 gives the total energy of a system that can be converted into heat. However, the second law of thermodynamics states that energy conversion is never 100%; some energy is always lost as entropy.

The Gibbs free energy formula is

$$\Delta <em> G </em>= \Delta H - T\Delta S$$

For octane combustion,

$$\Delta G = -10{,}149.038 \text{ kJ} - T\ (968.04 \text{ J/K})$$

Before we can finish the calculation, we must do two things. First, we need to pick a temperature. There isn't one "right" temperature, so let's use 25 °C (298 K).

$$\Delta G = -10{,}149.038 \text{ kJ} - (298 \text{ K})(968.04 \text{ J/K})$$

$$\Delta G = -10{,}149.038 \text{ kJ} - 288{,}475.92 \text{ J}$$

The second thing to do is change joules to kilojoules:

$$\Delta G = -10{,}149.038 \text{ kJ} - (288{,}475.92 \text{ J})(1 \text{ kJ}/1{,}000 \text{ J})$$

$$\Delta G = -10{,}149.038 \text{ kJ} - 288.47592 \text{ kJ}$$

$$\Delta G = -10{,}438 \text{ kJ}$$

spontaneous: a process that, once started, continues without input of energy
nonspontaneous: a process that requires a constant energy input

The sign of ΔG is negative, indicating that the reaction is spontaneous. If ΔG is positive, the reaction is nonspontaneous. Spontaneous reactions liberate energy that can be converted into work. Nonspontaneous reactions require an energy supply. Depending on its sign, ΔG tells us the maximum energy available for work, or the minimum energy that must be supplied for the reaction to occur. Chemical reactions that have positive ΔG are storing energy in chemical bonds.

Comparing Different Fuels

We can calculate ΔG for the combustion of ethanol, following the steps previously described. The ΔH_f^0 for $CH_3CH_2OH(g)$ is −235.1 kJ/mole, S^0 is −282.7 J/moleK, and the balanced chemical equation is

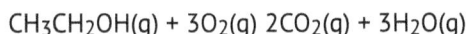

$$CH_3CH_2OH(g) + 3O_2(g) \rightarrow 2CO_2(g) + 3H_2O(g)$$

The ΔG for ethanol at 25 °C (298 K) is −1,474.3 kJ.

I've chosen octane because it is a very reasonable "model" compound to represent gasoline. There is no "gasoline" molecule: Gasoline is a mixture of hundreds of different molecules. Some of these molecules release more energy when burned, some release less. In the United States, ethanol is added to gasoline, so it is also an important fuel.

Clearly, octane releases more energy than ethanol does, but simply comparing ΔG values isn't a fair comparison. The ΔG for ethanol was calculated based on burning 1 mole of ethanol. When we burned octane, 2 moles of octane were burned. Having a higher value of ΔG for the octane combustion is partly due to burning more octane. We don't generally buy fuel by the mole, but by the gallon, so a more reasonable comparison would be this: "How much energy per gallon of fuel do we get from octane and ethanol?"

We need to do some calculations. First, we find how many moles of each fuel are present in 1 gallon of the fuel. Since 1 gallon = 3,785 milliliters (mL), multiplying the density value of each fuel by 3,785 mL gives the mass of each fuel in 1 gallon. For octane, 1 gallon is 2,661 grams; for ethanol, 1 gallon is 2,986.4 g. Using the formula weights for octane and ethanol, we calculate the moles of each fuel present. Finally, using the ΔG values and the number of moles, we calculate the energy content of each fuel. These values are shown in Table 7.1.

Octane releases ~27% more energy/gallon than ethanol. Simply comparing the ΔG's would indicate that octane has seven times more energy than alcohol. This is misleading. To get the equivalent energy of 1 gallon of octane, we need to burn 1.27 gallons of alcohol.

Table 7.1. Energy per Gallon of Fuel

	Octane	Ethanol
ΔG	−10,438 kJ	−1,474.3 kJ
Moles	2	1
Energy released per mole of fuel	−5,219 kJ	−1,474.3 kJ
1 gallon =	3,785 mL	3,785 mL
Density	0.703 g/mL	0.789 g/mL
Mass, 1 gallon	2,661 g	2,986.4 g
Formula weight	114.23	46.041
Moles in 1 gallon	23.295	64.864
Energy released per gallon	−121,577 kJ	−95,629 kJ

What about real gasoline? Work performed by the National Institute of Standards and Technology (NIST) in 1994 resulted in the Gasoline Gallon Equivalent (GGE)—a way of directly comparing the energy content of various fuels and fuel mixtures.

Gasoline provides 114,000 BTU (British Thermal Units, an alternate unit for energy) per gallon. The most common gasoline blend sold in the United States is E10, gasoline containing 10% ethanol. E10 provides 111,836 BTU/gallon, or 98.7% of the energy in "pure" gasoline. Ethanol provides 76,100 BTU/gallon, which is 66.7% of the energy in "pure" gasoline. It takes 1.5 gallons of ethanol to provide the same energy as 1 gallon of "pure" gasoline.

Another comparison would be the amount of greenhouse gas produced by each fuel. According to the Energy Information Administration (EIA), 1 gallon of pure gasoline produces 19.64 pounds of CO_2, 1 gallon of E10 produces 18.95 pounds of CO_2, and 1 gallon of pure ethanol produces 12.72 pounds of CO_2. In terms of energy provided per pound of CO_2 released, pure gasoline provides 5,804.5 BTU/lb, E10 provides 5,901.6 BTU/lb, and pure ethanol provides 5,982.7 BTU/lb.

Which fuel is better? The answer to this question, like most questions, depends on what is most important to you.

List of Key Takeaways From This Chapter

- The ΔH^0 (enthalpy) is the maximum energy released by the reaction as heat.
- If ΔH^0 is negative, the reaction is exothermic. If ΔH^0 is positive, the reaction is endothermic.
- The ΔS (entropy) is the change in "randomness" of the reaction. Temperature affects the absolute amount of entropy change.
- The ΔG (Gibbs free energy) combines the heat released (enthalpy) and the randomness change (entropy).
- If ΔG is negative, the reaction is spontaneous and energy is released. If ΔG is positive, the reaction is nonspontaneous and energy is absorbed.
- For realistic comparisons of energy released by different fuels, various factors must be included. A simple comparison of ΔG is not sufficient.

Chapter 7 Exercises

For the following chemical reactions, calculate the enthalpy of reaction (ΔH^0_{rxn}), the entropy of the reaction (ΔS^0), and the Gibbs free energy (ΔG^0). For calculating Gibbs free energy, use 25 °C (298 K). Tell whether the reaction is spontaneous or nonspontaneous. Note: You might have to balance the chemical reaction before calculating your values.

Values for ΔH^0 and ΔS are given in the table following the exercises.

1. $BaO(s) + 2HCl(aq) \rightarrow BaCl_2(s) + H_2O(l)$
2. $C_6H_6(l, benzene) + O_2(g) \rightarrow CO_2(g) + H_2O(g)$
3. $6H_2O(l) + 6C(s, graphite) \rightarrow 3C_2H_4(g, ethylene) + 3O_2(g)$
4. $NaOH(aq) + HCl(aq) \rightarrow NaCl(aq) + H_2O(l)$

5. $CaC_2(s) + H_2O(l)$ $C_2H_2(g, \text{acetylene}) + Ca(OH)_2(s)$
6. $2H_2O(l)$ $2H_2(g) + O_2(g)$
7. $2NiO(s)$ $2Ni(s) + O_2(g)$
8. $2CH_3OH(l) + 3O_2(g)$ $2CO_2(g) + 4H_2O(g)$

Substance	ΔH^0 kJ/mole	S^0 J/mol-K
$BaCl_2(s)$	−858.6	123.681
$BaO(s)$	−553.5	70.42
$CaC_2(s)$	−59.8	69.96
$Ca(OH)_2(s)$	−986.09	83.39
$C(s, \text{graphite})$	0	5.74
$C_2H_2(g, \text{acetylene})$	226.73	200.94
$C_2H_4(g, \text{ethylene})$	52.26	219.56
$C_6H_6(l, \text{benzene})$	49.03	172.8
$CH_3OH(l)$	−238.66	126.8
$CO_2(g)$	−393.509	213.74
$H_2(g)$	0	130.684
$HCl(aq)$	−167.159	56.5
$H_2O(l)$	−285.830	69.91
$NaCl(aq)$	−407.27	115.5
$NaOH(aq)$	−470.144	48.1
$Ni(s)$	0	29.87
$NiO(s)$	−239.7	37.99
$O_2(g)$	0	205.138

Answers

1. BaO(s) + 2HCl(aq) BaCl$_2$(s) + H$_2$O(l)

    $$\Delta H^0 = (-858.6 + (-285.830)) - (-553.5 + (2 \times -167.159))$$
    $$= -1144.43\text{kJ} - (-887.818\text{kJ})$$
    $$= -256.612\text{kJ}$$

    $$\Delta S^0 = (123.68\text{J/K} + 69.91\text{J/K}) - (70.42\text{J/K} + 2 \times 56.5\text{J/K})$$
    $$= 193.59\text{J/K} - 183.42\text{J/K}$$
    $$= 10.17\text{J/K}$$

    $$\Delta G^0 = -256.612\text{kJ} - (298\text{K} \times 10.17\text{J/K} \times \tfrac{1\text{kJ}}{1000\text{J}})$$
    $$= -256.612\text{kJ} - 3.03066\text{kJ}$$
    $$= -259.6\text{kJ}$$

Reaction is spontaneous.

2. 2C$_6$H$_6$(l, benzene) + 15O$_2$(g) 12CO$_2$(g) + 6H$_2$O(g)

    $$\Delta H^0 = ((-393.509 \times 12) + (-241.818 \times 6)) - ((2 \times 49.03) + (15 \times 0))$$
    $$= -6173.016\text{kJ} - (98.06\text{kJ})$$
    $$= -6271.076\text{kJ}$$

    $$\Delta S^0 = ((213.74\text{J/K} \times 12) + (188.825\text{J/K} \times 6)) - ((2 \times 172.8\text{J/K}) + (15 \times 205.138\text{J/K}))$$
    $$= 3697.83\text{J/K} - 3422.67\text{J/K}$$
    $$= 275.16\text{J/K}$$

    $$\Delta G^0 = -6271.076\text{kJ} - (298\text{K} \times 275.16\text{J/K} \times \tfrac{1\text{kJ}}{1000\text{J}})$$
    $$= -6271.076\text{kJ} - 81.99768\text{kJ}$$
    $$= -6353.07\text{kJ}$$

Reaction is spontaneous.

3. $6H_2O(l) + 6C(s, graphite) \; 3C_2H_4(g, ethylene) + 3O_2(g)$

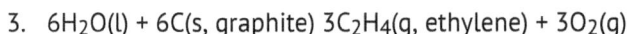

$$\Delta H^0 = ((52.26 \times 3) + (0 \times 3)) - ((-285.83 \times 6) + (6 \times 0))$$
$$= 156.78\text{kJ} - (-1714.98\text{kJ})$$
$$= +1871.76\text{kJ (Notethesign!!!)}$$

$$\Delta S^0 = ((219.56\text{J/K} \times 3) + (205.138\text{J/K} \times 3)) - ((6 \times 69.91\text{J/K}) +$$
$$(6 \times 5.74\text{J/K}))$$
$$= 1274.094\text{J/K} - 453.9\text{J/K}$$
$$= 820.194\text{J/K}$$

$$\Delta G^0 = +1871.76\text{kJ} - (298\text{K} \times 820.194\text{J/K} \times \tfrac{1\text{kJ}}{1000\text{J}})$$
$$= +1871.76\text{kJ} - 244.417812\text{kJ}$$
$$= +1627.34\text{kJ}$$

Reaction is nonspontaneous as written.

4. $NaOH(aq) + HCl(aq) \; NaCl(aq) + H_2O(l)$

$$\Delta H^0 = (-407.27 + (-285.830)) - (-470.114 + (-167.159))$$
$$= -693.1\text{kJ} - (-637.273\text{kJ})$$
$$= -55.827\text{kJ}$$

$$\Delta S^0 = (115.5\text{J/K} + 69.91\text{J/K}) - (48.1\text{J/K} + 56.5\text{J/K})$$
$$= 185.41\text{J/K} - 104.6\text{J/K}$$
$$= 80.81\text{J/K}$$

$$\Delta G^0 = -55.827\text{kJ} - (298\text{K} \times 80.81\text{J/K} \times \tfrac{1\text{kJ}}{1000\text{J}})$$
$$= -55.827\text{kJ} - 24.08138\text{kJ}$$
$$= -79.91\text{kJ}$$

Reaction is spontaneous.

5. $CaC_2(s) + 2H_2O(l) \rightarrow C_2H_2(g, \text{acetylene}) + Ca(OH)_2(s)$

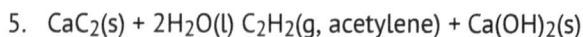

$$\Delta H^0 = (226.73 + (-986.09)) - (-59.8 + (-285.830 \times 2))$$
$$= -759.36\text{kJ} - (-631.46\text{kJ})$$
$$= -127.9\text{kJ}$$

$$\Delta S^0 = (200.94\text{J/K} + 83.39\text{J/K}) - (69.96\text{J/K} + (2 \times 69.91\text{J/K}))$$
$$= 284.33\text{J/K} - 209.78\text{J/K}$$
$$= 74.55\text{J/K}$$

$$\Delta G^0 = -127.9\text{kJ} - (298\text{K} \times 74.55\text{J/K} \times \tfrac{1\text{kJ}}{1000\text{J}})$$
$$= -127.9\text{kJ} - 22.2159\text{kJ}$$
$$= -150.16\text{kJ}$$

Reaction is spontaneous.

6. $2H_2O(l) \rightarrow 2H_2(g) + O_2(g)$

$$\Delta H^0 = ((2 \times 0) + 0) - (-285.830 \times 2)$$
$$= 0\text{kJ} - (-571.66\text{kJ})$$
$$= +571.66\text{kJ}(\text{Notesign!!})$$

$$\Delta S^0 = ((2 \times 130.684\text{J/K}) + 205.138\text{J/K}) - (2 \times 69.91\text{J/K})$$
$$= 466.506\text{J/K} - 139.82\text{J/K}$$
$$= 326.686\text{J/K}$$

$$\Delta G^0 = +571.66\text{kJ} - (298\text{K} \times 326.686\text{J/K} \times \tfrac{1\text{kJ}}{1000\text{J}})$$
$$= +571.66\text{kJ} - 97.352428\text{kJ}$$
$$= +474.31\text{kJ}$$

Reaction is nonspontaneous as written.

7. 2NiO(s) 2Ni(s) + O₂(g)

$$\Delta H^0 = ((2 \times 0) + 0) - (-239.7 \times 2)$$
$$= 0\text{kJ} - (-479.4\text{kJ})$$
$$= +479.4\text{kJ}(\text{Notesign!!})$$

$$\Delta S^0 = ((2 \times 29.87\text{J/K}) + 205.138\text{J/K}) - (2 \times 37.99\text{J/K})$$
$$= 264.878\text{J/K} - 75.98\text{J/K}$$
$$= 188.898\text{J/K}$$

$$\Delta G^0 = +479.4\text{kJ} - (298\text{K} \times 188.898\text{J/K} \times \tfrac{1\text{kJ}}{1000\text{J}})$$
$$= +479.4\text{kJ} - 56.291604\text{kJ}$$
$$= +423.11\text{kJ}$$

Reaction is nonspontaneous as written.

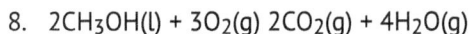

8. 2CH₃OH(l) + 3O₂(g) 2CO₂(g) + 4H₂O(g)

$$\Delta H^0 = ((2 \times -393.509) + (4 \times -241.818)) - ((-238.66 \times 2) + (3 \times 0))$$
$$= -1754.29\text{kJ} - (-477.32\text{kJ})$$
$$= -1276.97\text{kJ}$$

$$\Delta S^0 = ((2 \times 213.74\text{J/K}) + (4 \times 188.825\text{J/K})) -$$
$$((2 \times 126.8\text{J/K}) + (3 \times 205.138))$$
$$= 1182.78\text{J/K} - 869.014\text{J/K}$$
$$= 313.766\text{J/K}$$

$$\Delta G^0 = -1276.97\text{kJ} - (298\text{K} \times 313.766\text{J/K} \times \tfrac{1\text{kJ}}{1000\text{J}})$$
$$= -1276.97\text{kJ} - 93.502268\text{kJ}$$
$$= -1370.47\text{kJ}$$

Reaction is spontaneous as written.

Fossil Fuels Versus Renewable Fuels

> We know we'll run out of dead dinosaurs to mine for fuel and have to use sustainable energy eventually, so why not go renewable now and avoid increasing risk of climate catastrophe? Betting that science is wrong and oil companies are right is the dumbest experiment in history by far.
>
> Elon Musk, Founder, Tesla, Inc.

In 2020, the United States used a total of 92.94 quadrillion BTU of energy, from five sources. The largest source was petroleum (35% of total), followed by natural gas (34%), renewable energy sources (12%), coal (10%), and nuclear electric power (9%).

I'm not going to cover nuclear power, because nuclear processes are entirely different than chemical processes and outside the scope of this text. I will cover the other four sources in some detail.

(Please note: Due to fluctuations in costs, production, and demand, the values used in this section are constantly changing. Unless otherwise noted, all values are from 2020–2022, and are based on U.S. production and usage.)

Learning Objectives

This chapter will help students

- Distinguish between fossil fuels and renewable fuels
- Learn the common uses of fossil fuels and renewable fuels
- Compare the benefits and problems with each fossil fuel and renewable fuel
- Understand how each energy source is obtained and how it is typically used

Fossil Fuels

fossil fuel: an energy source formed in the Earth's crust over millions of years from decaying organic material

All **fossil fuels** were produced by chemical and geological processes taking millions of years, under conditions that aren't operating today. Necessarily, the total amount of fossil fuel in the world is finite, although no one is sure of the exact amount remaining. There are three common fossil fuels: coal, natural gas, and petroleum.

Coal

Coal has been an important fuel for centuries and has been used for heating buildings, refining metals, producing illuminating gas, and as fuel for transportation. In 2020–2022, coal accounted for 9.4 quadrillion BTU in the United States. About 90% of all coal is used to produce electricity. Another 10% is used for refining metal, producing concrete and paper, and producing synthetic fuels (synfuels). Table 8.1 summarizes the various types of coal.

Table 8.1. Various Types of Coal Used in the U.S.

Coal type	Carbon (%)	Heat value (BTU/lb)	U.S. production (%)
Anthracite	86–97	~14,000	1
Bituminous	45–86	~12,800	44
Sub-bituminous	35–45	~9,000	46
Lignite	25–35	~6,900	9

Data courtesy of U.S. EIA

The precursor to coal is peat, a soft material composed of partially decomposed plants and minerals. Geological processes involving heat, pressure, and time converted peat into lignite, and then successively into sub-bituminous, bituminous, and anthracite coals.

Coal is obtained using either surface mining methods or deep mining methods. Surface mining (strip mining) removes the topsoil and rock to uncover the coal. In deep mining, tunnels are dug into mountains or hillsides, or vertically into the ground, to access the coal. About 63% of U.S. coal production comes from surface mining. Five states (Wyoming, West Virginia, Pennsylvania, Illinois, and North Dakota) account for 72% of total surface mining production. Water that drains from coal mines may be contaminated with toxic substances, and federal law requires that this water be controlled, and the land reclaimed after mining.

Before the mined coal is shipped to consumers, it is crushed and extraneous noncoal material is removed. High-sulfur coals are washed with water or a chemical bath to remove sulfur. This washing liquid is collected in wastewater ponds known as "slurry" impoundments. Processing coal generates large amounts of waste; typically, 35%–40% of unprocessed coal entering a washing facility is left behind as waste.

When burned, coal produces a variety of hazardous substances. Any sulfur present produces sulfur dioxide gas (SO_2). Sulfur dioxide reacts with water to produce sulfurous acid (H_2SO_3), one component of acid rain. Nitrogen oxides, generally shown as NO_x, are another product of combustion. These react with water to form nitrous and nitric acids (HNO_2, HNO_3). Particulate matter, including soot, smoke, **fly ash**, and dust, contributes to smog, haze, and lung diseases. About 197,000 tons of particulate matter are released annually from burning coal.

fly ash: a finely divided residue left from coal combustion

In 2021 there were 240 coal-burning power plants in the United States, producing about 899 billion kilowatt-hours (kWh) of electricity. These plants emitted about 12 million tons of sulfur dioxide, which was 93% of all sulfur dioxide emitted by electric utilities in the U.S. The plants also emitted about

780,000 tons of NO_x, about 10% of total NO_x emissions. For the last several years, these emissions have been trending down.

Fly ash from burning coal contains hazardous metals such as lead, nickel, cadmium, and arsenic. These metals can leach out of the ash into groundwater and are hazardous to human health. The U.S. Environmental Protection Agency (EPA) estimates that 140 million tons of coal fly ash are produced annually. Most of this ash is stored at the power plant where it's used, or it is buried in landfills. About 43% is recycled into hydraulic cement.

In the U.S., coal demand peaked in 2007 and has declined nearly every year since then, mainly because many power plants have converted to natural gas.

Natural Gas

Natural gas was first used in Britain in 1785. It was produced by strongly heating coal without air and was used to light houses and streetlights. In 1836, Philadelphia opened the first municipally owned gas company in the U.S. Throughout the 19th century, natural gas was mostly used for lighting homes, businesses, and streetlamps. Once sufficient pipelines were built, the gas was also used for heating and cooking and as fuel to power industrial boilers. Today, about 99% of the natural gas used in the United States is produced domestically.

Modern natural gas comes from oil and gas wells, and it accounts for 34% (31.6 quadrillion BTU) of all energy used in the U.S. It supplies about half of residential and commercial energy and about 41% of industrial energy. About 38% of natural gas in the U.S. is used to produce electricity.

Natural gas is a mixture of simple compounds. Methane (CH_4) is the major component, with smaller amounts of ethane (C_2H_6), propane (C_3H_8), and butane (C_4H_{10}).

Natural gas generally comes from the same wells that produce petroleum, although some wells produce only natural gas. Natural gas processing starts at the well, where gas is separated from crude oil. Next, water is removed by absorption or adsorption methods. This reduces corrosion of pipelines and other gas-handling equipment. Acid gases, mostly hydrogen sulfide (H_2S) and carbon dioxide (CO_2), are removed, and finally nitrogen compounds are removed. The processed gas is then ready for delivery.

Combustion of natural gas produces carbon dioxide and water. Depending on combustion efficiency, trace amounts of carbon monoxide (CO) may be produced. Other substances such as nitrogen and sulfur oxides and soot may be produced. Natural gas is generally considered the cleanest-burning fuel available, making it much more attractive than coal.

Petroleum

Petroleum supplies about 35% of U.S. energy (32.5 quadrillion BTU). About 68% of all petroleum used in the U.S. is for transportation, while 26% is used in the industrial sector. Industrial uses include manufacturing artificial rubber, plastics, asphalt, and the petrochemicals that are used as starting materials for thousands of products.

Petroleum has been used for thousands of years, with ancient peoples using natural asphalt for construction and for paving streets. The Chinese used petroleum as a fuel in the fourth century B.C.E. and drilled the earliest known well in 347 C.E. In the ninth century, oil fields were discovered and exploited in Azerbaijan. From the ninth to the twelfth century, Arab and Persian chemists distilled petroleum to produce kerosene and other substances. In the U.S., the first intentionally drilled oil well was drilled in 1859 in Titusville, Pennsylvania.

In the U.S. there are currently ~937,000 active oil and gas wells, producing about 7 billion barrels of crude oil and petroleum products in 2021. U.S. production was 18.61 million barrels per day, or 20% of the world's production. Table 8.2 compares petroleum production in the leading oil-producing countries in 2021.

Table 8.2. Crude Oil Production in 2021 for Top 10 Oil-producing Countries

Country	Barrels/day (million)	Share of world total (%)
United States	18.61	20
Saudi Arabia	10.81	12
Russia	10.50	11
Canada	5.23	6
China	4.86	5
Iraq	4.16	4
United Arab Emirates	3.78	4
Brazil	3.77	4
Iran	3.01	3
Kuwait	2.75	3
Top 10 total	67.49	72
World total	93.86	

Data courtesy of U.S. EIA

Regardless of how much it produces in-country, the U.S. still imports a significant amount of petroleum (both crude oil and refined products). In 2021, the U.S. purchased 3.09 billion barrels of petroleum products from 73 countries. Crude oil accounted for 72% of these purchases. Most petroleum purchases were from Canada (51%), Mexico (8%), Russia (8%), Saudi Arabia (5%), and Columbia (2%).

In the same year, the U.S. exported 3.15 billion barrels of petroleum products to 176 countries and 4 U.S. territories. Crude oil accounted for 35% of these exports. The major purchasers were Mexico (13%), Canada (10%), India (7%), China (7%), and South Korea (6%). In 2021, the U.S. was a net petroleum exporter.

Petroleum processing is covered in more detail in the next chapter.

Renewable Energy

In 2021–2022, renewable energy sources accounted for 12% (11.59 quadrillion BTU) of energy consumption in the U.S., and that amount is steadily growing. **Renewable energy** is typically used for electricity generation, heating, and transportation. Not all renewable energy sources are chemically based, but all renewable sources can be used to produce electricity, and electricity from any source is important to our society. Table 8.3 summarizes the various types of renewable energy sources.

> **renewable energy**: energy from a source that is not depleted when used, such as wind or solar power

Table 8.3. Percentage Contribution of Renewable Energy Sources, U.S., 2021

Source	% of Renewable	Most important use
Wind	26	Electricity
Hydroelectric	22	Electricity
Wood	18	Heat
Biofuels	17	Transportation
Solar	11	Electricity, heat
Biomass waste	4	Electricity, heat
Geothermal	2	Heat, electricity, bathing

Data courtesy of U.S. EIA

Wind

Wind has been used as an energy source for centuries. The first known sailing ship was discovered in Egypt and is about 5,500 years old. The first practical windmills, used to pump water or grind grain, were in Persia (modern-day Iran) and date to the seventh to ninth centuries C.E.

In 2020–2022, wind accounted for 3.01 quadrillion BTU in the United States. Wind is highly variable throughout the day and from season to season. Generally, wind turbines are located where the wind blows relatively constantly at 9–13 mph. Locations with smooth, rounded hilltops, open plains, open water, or mountain passes that can funnel the wind are choice sites. Generally, the wind velocity is higher above ground level, and many wind turbines are 500–900 feet tall. Currently, 42 states have utility-scale wind power projects (Figure 8.1). The U.S. produced 21% of all wind electricity worldwide. China (30%), Germany (8%), the United Kingdom (5%), and India (4%) are other world leaders.

U.S. utility-scale wind electricity generation by state, 2021

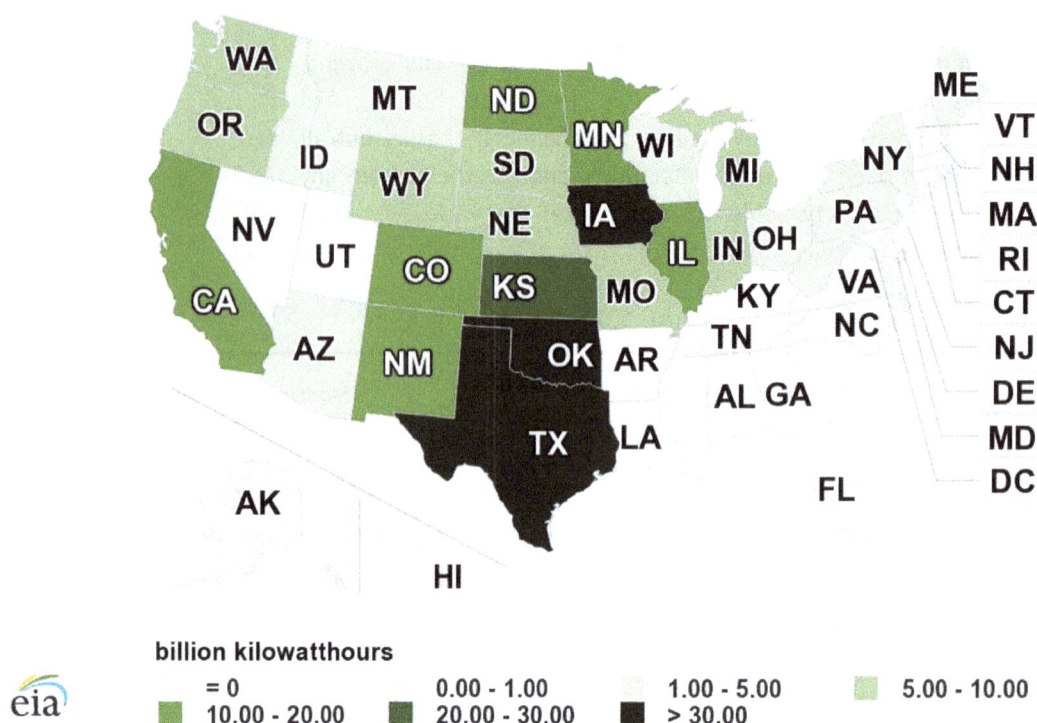

billion kilowatthours

| = 0 | 0.00 - 1.00 | 1.00 - 5.00 | 5.00 - 10.00 |
| 10.00 - 20.00 | 20.00 - 30.00 | > 30.00 | |

Figure 8.1   U.S. Utility-scale Wind Electricity Generation by State, 2021

United States Energy Information Administration, https://www.eia.gov/energyexplained/wind/where-wind-power-is-harnessed.php, 2021.

Hydroelectric

The first hydroelectric plant was built in Craigside, Northumberland, England, by William Armstrong (1810–1900), an English engineer, and was used to power an arc lamp in Armstrong's art gallery. The first U.S. industrial plant was built in 1880 in Grand Rapids, Michigan, to power arc lamps at the Wolverine Chair Factory. The first plant to sell electricity was opened on Fox River, Appleton, Wisconsin, on September 30, 1882.

There are about 1,450 conventional and 40 pumped storage plants in the U.S., accounting for 2.55 quadrillion BTU. Conventional hydroelectric plants use a dam to create a reservoir and direct the water through a turbine, which generates the electricity. Pumped storage plants use two reservoirs at different heights; water flowing from the higher reservoir is directed through the power-generating turbine. Pumped storage plants store energy until it is needed.

The oldest U.S. plant still operating is the Whiting Plant, in Whiting, Wisconsin, a 4-megawatt (MW) plant built in 1891. Most hydroelectric plants are large dams built on major rivers, and they were built by various federal agencies before the mid-1970s. The largest U.S. plant is the Grand Coulee Dam on the Columbia River in Washington, producing 6,785 MW. Figure 8.2 shows the distribution of hydroelectric power in the United States.

Hydroelectricity generation by state in 2021

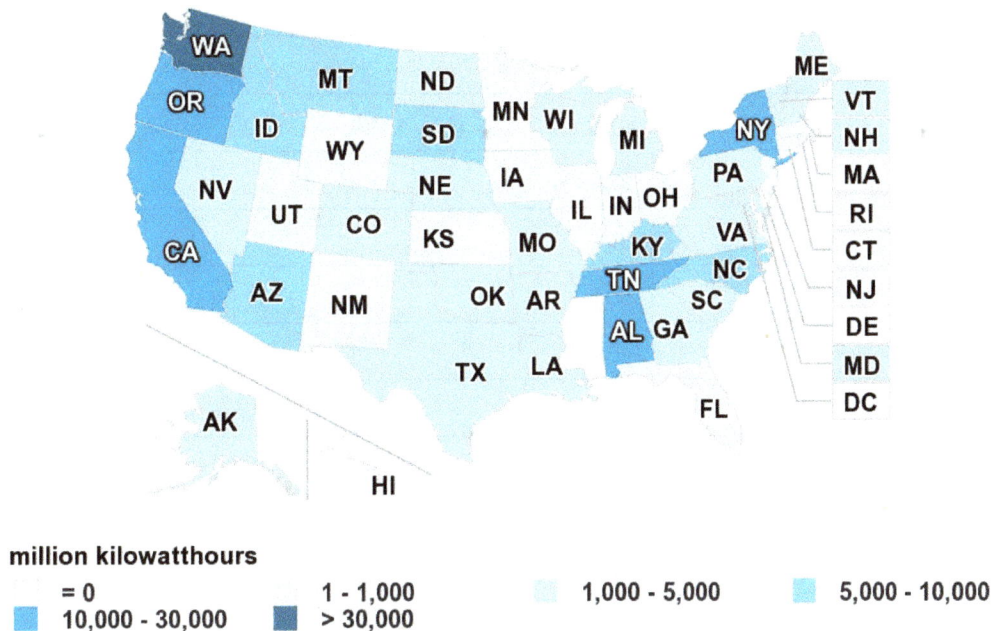

million kilowatthours
- = 0
- 1 - 1,000
- 1,000 - 5,000
- 5,000 - 10,000
- 10,000 - 30,000
- > 30,000

Source: U.S. Energy Information Administration, *Electric Power Monthly*, Table 1.10.B, February 2022, preliminary data

Figure 8.2   Hydroelectricity Generation by State, 2021

United States Energy Information Administration, https://www.eia.gov/energyexplained/hydropower/where-hydropower-is-generated.php, 2021.

Of the 50 states, Washington (27%), California (13%), Oregon (10%), New York (6%), and Alabama (4%) are the top hydroelectric producers. All states except Mississippi produce some hydroelectricity. Most dams were built for flood control, with only a small percentage specifically built for electricity production. The U.S. Department of Energy (DOE) estimates that 12,000 MW of electricity generation capacity is available from flood control dams.

Worldwide hydroelectric capacity is 1,308 gigawatts (GW). Table 8.4 shows the top 10 countries that produce hydroelectric power, accounting for 68% of worldwide hydroelectric capacity.

Table 8.4. Top 10 Producers of Hydroelectric Power

Country	Capacity (GW)	%
China	356.4	27
Brazil	109.1	8
U.S.	102.8	8
Canada	81.4	6
India	50.1	4
Japan	49.9	4
Russia	49.9	4
Norway	32.7	3
Turkey	28.5	2
France	25.6	2
Top 10 total	886.4	68
Global total	1,308	

Data courtesy of U.S. EIA

Wood

In 2021–2022, wood and wood waste accounted for 2.14 quadrillion BTU. About 65% of this energy is used by the lumber and paper industries to produce steam, heat, and locally generated electricity. Residential use is the next largest consumer, with 22% of wood energy used in fireplaces and wood-burning stoves. Wood energy represents 4% of all residential-sector end-use energy consumption and 2.2% of total residential energy consumption. About 12.5 million households (~11% of all U.S. households) use wood, mostly for space heating. About 3.5 million households use wood as their main heating fuel.

Some coal-burning electric plants burn wood and wood chips with the coal, to reduce their total sulfur emissions. Most of the commercial sector's wood use is for space heating. Electricity and commercial-sector wood use accounts for 13% of total wood energy.

Globally, about 7% of total energy consumption is from wood, with developing countries using 76% of the total. Figure 8.3 shows the worldwide distribution of wood fuel usage.

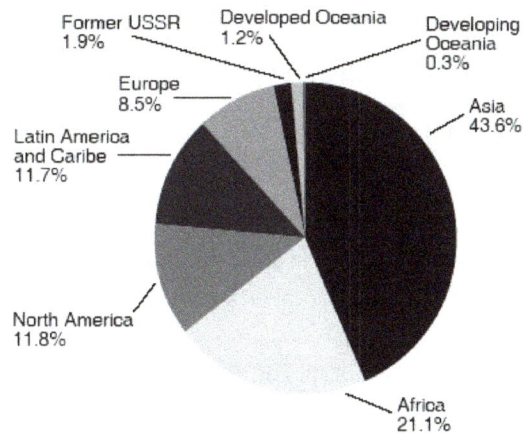

Source: FAO's Wood Energy Information System.

Figure 8.3   Distribution of Wood Energy by Region

M.A. Trossero/FAO, https://www.fao.org/3/y4450e/y4450e02.htm.
Copyright © 2002 by United Nations.

Biofuels

Biofuels accounted for about 1.97 quadrillion BTU. Four major categories are ethanol, biodiesel, renewable diesel, and other biofuels (such as renewable jet fuel, renewable heating oil).

Ethanol is an alcohol produced by fermentation of sugars from fruits or grains and is blended with gasoline. In the U.S., nearly all ethanol used for fuel comes from corn. Almost all gasoline sold in the U.S. contains 10% ethanol by volume. Ethanol has the largest share of biofuel production (85%) and consumption (82%) in the U.S.

> **biofuels:** liquid fuels commonly used for transportation

Biodiesel is usually blended with petroleum diesel and has the second largest share of biofuel production (11%) and consumption (12%). Biodiesel was one of the first biofuels developed. Rudolf Diesel (1858–1913), the German inventor and mechanical engineer who developed the diesel engine in 1897, experimented with vegetable oil as a fuel for his engine.

Biodiesel is produced by chemically converting vegetable oils and fats using a process called *transesterification*. In this reaction, fatty acid molecules are reacted with methanol and sodium hydroxide to produce glycerin and fatty acid methyl esters. The most common sources of oils for biodiesel are soybean oil (57%), corn oil (14%), recycled cooking oils (11%), canola oil (10%), and animal fats (8%).

Renewable diesel and other biofuels are produced from almost any kind of biomass, using a variety of chemical processes such as hydrotreating, gasification, and pyrolysis. To produce renewable diesel requires a hydrogenation process instead of transesterification. In hydrogenation, extra hydrogen is added to the triglyceride molecules, eventually producing long molecules composed of only carbon and hydrogen atoms. These hydrocarbons are chemically equivalent to petroleum diesel. Unlike biodiesel, renewable diesel can be used as a direct replacement for diesel—it doesn't have to be blended with petroleum diesel.

Solar

In 1830, during an expedition to Africa, the British astronomer John Herschel (1792–1871) used a solar oven to cook food, the first recorded instance of the direct use of solar energy. There are many technologies for converting sunlight, including

- Thermal energy systems to heat water in homes, buildings, greenhouses, and other structures
- Systems to heat fluids to high temperatures in solar-thermal power plants
- Conversion of sunlight into electricity by photovoltaic (PV) systems

Solar energy by itself does not produce pollutants or greenhouse gases and has minimal environmental impact. However, like any other large-scale industrial process, some negative factors need to be considered.

The two biggest problems with solar energy are availability and concentration. Sunlight is not constant; it varies with time of day, season, and local weather conditions. The amount of sunlight per square foot of land is relatively low, so large areas are needed to collect significant amounts of sunlight. Figure 8.4 shows utility-scale solar generation of electricity (greater than 1 MW) in the U.S.

Utility-scale solar electricity generation by state, 2021

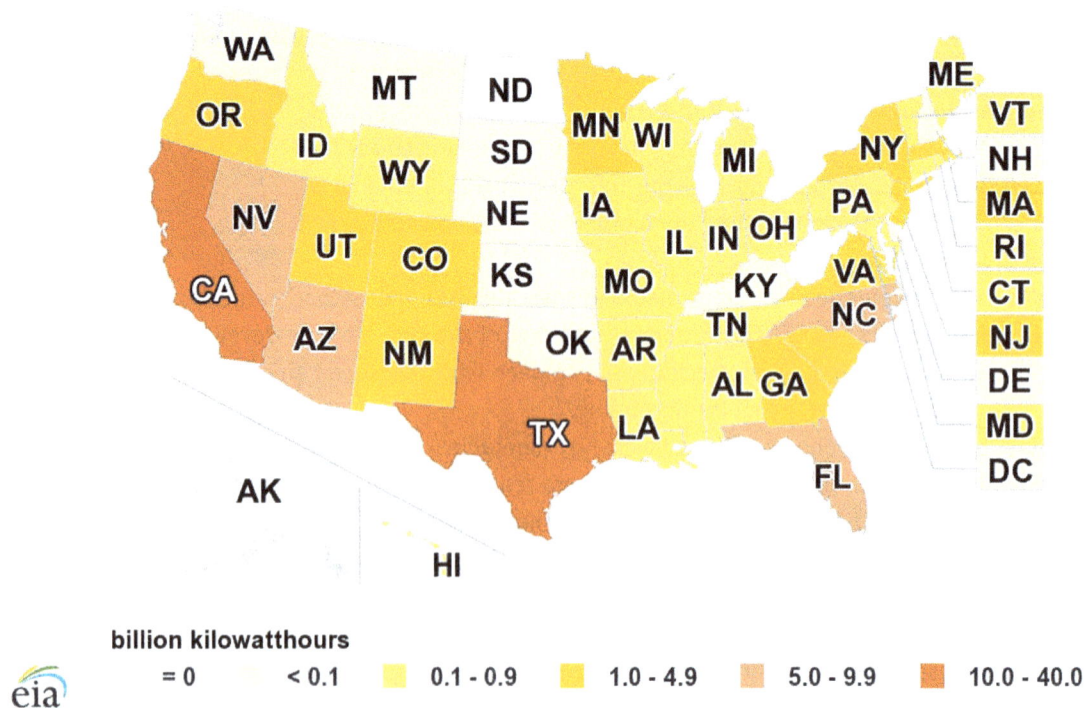

Figure 8.4   Utility-scale Solar Electricity Generation by State, 2021

United States Energy Information Administration, https://www.eia.gov/energyexplained/solar/where-solar-is-found.php, 2021.

In 2020–2021, solar energy provided 1.27 quadrillion BTU in the U.S. It accounted for 3% of electricity generated in 2020 and is projected to reach 20% by 2050. By then, about 80% of this electricity will be from utility-scale generation (1 MW or greater).

Worldwide in 2020, solar energy produced 842 billion kWh of electricity. China (32%), the U.S. (16%), Japan (9%), India (7%), and Germany (6%) accounted for 70% of solar-produced electricity.

The most common technologies using solar energy are **photovoltaic (PV)**, water heating systems, and concentrated solar power.

A PV cell (Figure 8.5) is made from semiconductor materials. When the semiconductor absorbs enough light energy, electrons are dislodged from the atoms. The electrons migrate to the surface of the cell, and from there through the electric circuit. The efficiency of current cells is about 20% for very-high-quality cells. Some experimental cells for space satellites have achieved 50% efficiency.

photovoltaic (PV): the conversion of light into electricity

Inside a photovoltaic cell

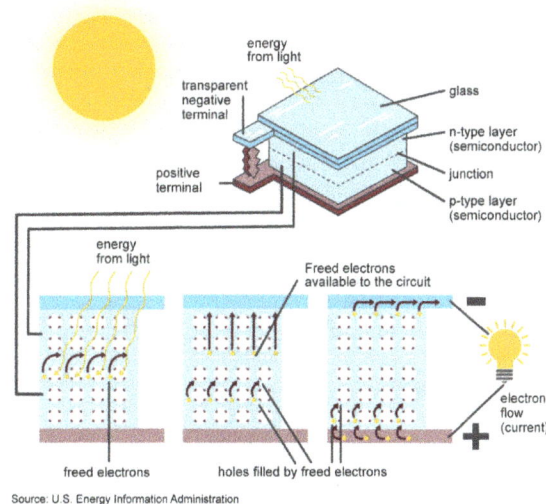

Source: U.S. Energy Information Administration

Figure 8.5   Inside a Photovoltaic Cell

United States Energy Information Administration, https://www.eia.gov/energyexplained/solar/photovoltaics-and-electricity.php, 2021.

Cells vary in size from about 0.5 inches to 4 inches in width. Typically, each cell produces 1 to 2 watts—enough power to run a calculator or a wristwatch. Individual cells are arranged into larger panels or modules, with the number of cells determining the power produced by the panel.

PV cells produce direct current (DC) electricity—the same kind of current produced by ordinary batteries. Nearly all electricity in the U.S. is supplied as alternating current (AC), so an inverter is used to convert DC into AC. Large photovoltaic systems provide electricity to pump water, operate communication equipment, supply electricity for single homes and businesses, or form larger arrays supplying electricity to cities. PV systems can supply electricity to areas where conventional electric service doesn't exist or has been damaged or destroyed. They can supply electricity to existing electric power grids. PV arrays can be installed relatively quickly, can be almost any size, and have minimal environmental effects when located on the roofs of buildings.

However, manufacturing PV cells requires heavy metals such as gallium and arsenic, which are toxic. Mining these ores creates significant environmental problems. Other hazardous materials that are used include hydrofluoric and sulfuric acids, and manufacturing the cells produces silicon dust, which causes inflammation and scarring of the lungs. Recycling PV cells is another major challenge that must be addressed in the near future.

A solar water-heating system uses a storage tank, and solar collectors mounted on the roof of a building (Figure 8.6). Sunlight heats water in the solar collector, and the water is stored in the tank. In climates where freezing temperatures are common, a different fluid is used in place of water. The heated fluid is passed through a heat exchanger to heat air, which is circulated through the building, or uses the hot water directly. If the system includes a pump and controller, then it is an "active" system. Solar heating systems require a backup system for cloudy days.

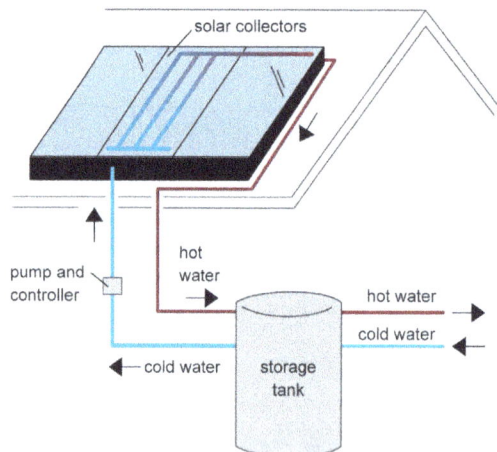

Basic components of a solar water heating system

Note: This is a simplified diagram of a drainback-type solar water heating system.
Source: U.S. Energy Information Administration

Figure 8.6   Basic Components of a Solar Water-heating System

United States Energy Information Administration, https://www.eia.gov/energyexplained/solar/solar-thermal-collectors.php, 2021.

A concentrated solar power system is a utility-scale producer of electricity (Figure 8.7). It is a huge array of mirrors that reflect and focus sunlight on a central collection tower, converting the light into heat. The heat produces electricity either by a steam turbine or by a heat engine driving a generator. Depending on specific design details, these systems produce from 80 MW to 200 MW of power.

Figure 8.7   Concentrated Solar Power Array

National Renewable Energy Laboratory/United States Department of Energy,
https://commons.wikimedia.org/wiki/File:Solar_two.jpg.

There are about a dozen concentrated solar power systems in the U.S. These systems require large land areas—typically 5 to 10 acres per MW of capacity. They must be located where sunlight is available throughout the year and they must have access to existing electric transmission grids. In the U.S., the most reasonable location is the Southwest, ranging from west Texas to southern California.

Biomass Waste

Simply put, trash accounted for about 0.46 quadrillion BTU. The U.S. produces about 292 million tons of trash annually, and approximately 12% of it is burned in waste-to-energy plants, producing steam and electricity. Paper, cardboard, food waste, grass clippings, wood, leather, plastics, and other combustible materials compose about 85% of trash. Noncombustible materials, like glass and metal, must be removed before the waste can be burned.

There are about 71 waste-to-energy plants in the U.S., located in 20 states. Most of these plants are in Florida, which has 20% of the total, and along the East Coast from Maryland to Maine (Figure 8.8). The plants produce about 2.3 GW of electricity. Leading countries in the number of waste-to-energy plants are Sweden, Denmark, the United Kingdom, Norway, Germany, the U.S., and the United Arab Emirates.

Municipal solid waste-to-energy plants with electricity generation capacity (2015)

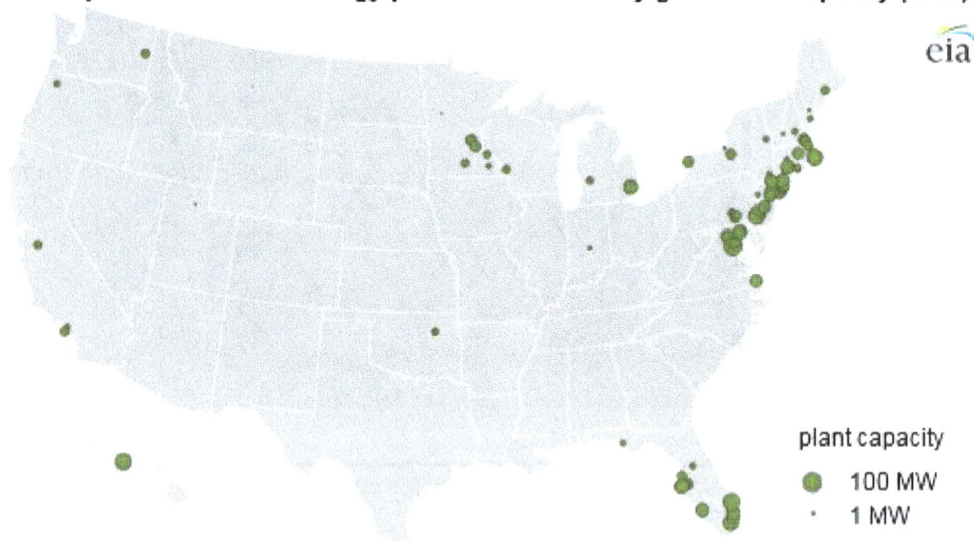

Figure 8.8   Municipal Waste-to-Energy Plants with Electricity-Generation Capacity, 2015

United States Energy Information Administration, https://www.eia.gov/todayinenergy/detail.php?id=25732, 2016.

Geothermal

Geothermal energy comes from heat produced within the Earth caused by radioactive disintegration. The most common uses of geothermal energy are for bathing (hot springs and spas), heating buildings, and producing electricity. Geothermal energy accounts for 0.23 quadrillion BTU.

Geothermal energy is clean, sustainable, abundant, and reliable. However, not every area has hydrothermal sources located near the surface. Withdrawing large amounts of hot water from hydrothermal sources has triggered earthquakes. Geothermal systems have higher initial costs than their competitors. A geothermal electric system may cost 2,500 per kWh, versus 1,000 per kWh for a natural gas system. Potential environmental issues include the release of dissolved gases, such as CO_2, and the presence of toxic metals such as chromium, vanadium, cobalt, and nickel.

Currently, three types of geothermal energy systems are used in the U.S. Direct-use systems take hot water from sources near the Earth's surface and pump the hot water directly into buildings. These systems can be used by individual homes, factories, or commercial buildings, and they can be part of larger municipal or district heating systems.

The oldest geothermal district heating system is probably in Chaides-Aigues in France, where 82 °C geothermal water was used to heat homes in the 14th century. In the United States, the oldest and largest system, which started in 1892, is in Boise, Idaho. Reykjavik, Iceland, has the world's largest municipal geothermal heating service. It supplies about 750 MW of thermal power, with 15.8 billion gallons of hot water flowing through the system annually.

Geothermal electricity generation requires water or steam between 150 °C and 370 °C. These systems are built close to large geothermal reservoirs. Figure 8.9 shows the distribution of geothermal electricity generation in the U.S. About 0.4% of utility-scale electricity, 16 billion kWh, comes from geothermal generation. Table 8.5 summarizes geothermal electricity generation for the U.S.

State rankings for geothermal electricity generation, 2021

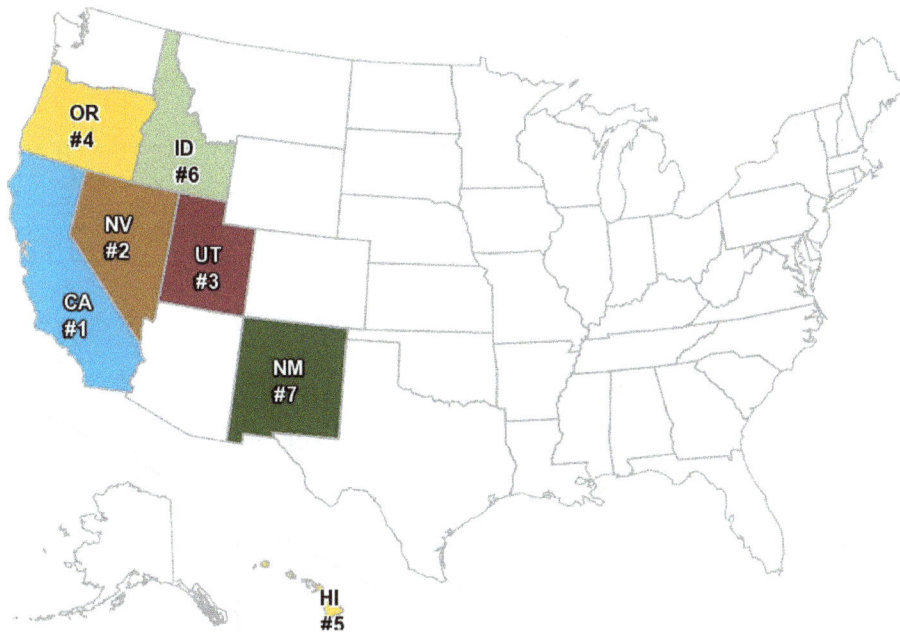

Source: U.S. Energy Information Administration, *Electric Power Monthly*, Table 16.1.B, February 2022, preliminary data

Figure 8.9   State Rankings for Geothermal Electricity Generation, 2021

United States Energy Information Administration, https://www.eia.gov/energyexplained/geothermal/use-of-geothermal-energy.php, 2022.

Table 8.5. Geothermal Electricity Production by State

State	kWh produced (billion)	U.S. total (%)	State total (%)
California	11.28	70.5	5.8
Nevada	3.87	24.2	9.4
Utah	0.35	2.2	0.8
Oregon	0.21	1.3	2.2
Hawaii	0.16	1.0	0.3
Idaho	0.08	0.5	0.5
New Mexico	0.05	0.3	0.1

Data courtesy of U.S. EIA

In 2019, 27 countries, including the U.S., generated 15,854 MW of electricity from geothermal sources. Table 8.6 shows the leading producers of geothermal electricity.

Table 8. 6 . World Leaders in Geothermally Generated Electricity

Country	Electricity produced (MW)	World total (%)
U.S.	3,639	23.0
Indonesia	1,984	12.5
Philippines	1,868	11.8
Turkey	1,347	8.5
New Zealand	1,005	6.3
Mexico	951	6.0
Italy	944	6.0
Iceland	755	4.8
Japan	601	3.8
Top 9 combined	13,904	82.6

Geothermal heat pumps rely on a very simple phenomenon: Ground temperature below the frost line is relatively constant at 50 °F–60 °F. In most U.S. locations, the frost line is 3–4 feet below the surface. A geothermal heat pump uses plastic pipe buried below the frost line. Water is circulated through the pipe and into a heat exchanger in the building. In the summer, the circulating water is cooler than the outside temperature and is used for air conditioning. In the winter, the water is warmer than the outside temperature and is used for heating. Figure 8.10 shows one example of a closed-loop geothermal heat pump system.

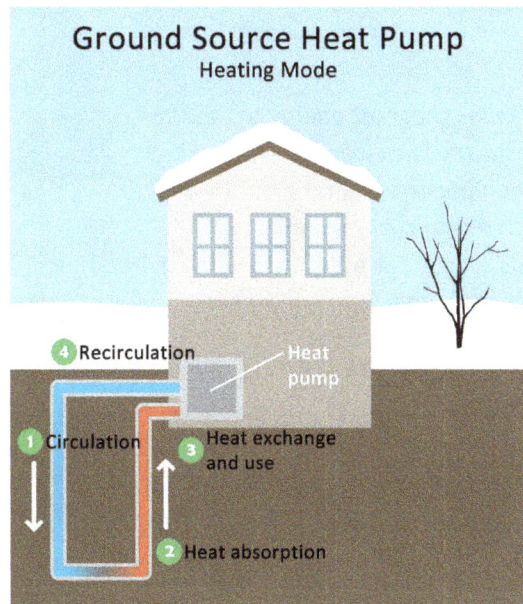

Ground Source Heat Pump
Heating Mode

4 Recirculation

Heat pump

1 Circulation

3 Heat exchange and use

2 Heat absorption

Figure 8.10   Closed-loop Geothermal Heat Pump

United States Environmental Protection Agency,
https://www.epa.gov/rhc/geothermal-heating-and-
cooling-technologies, 2022.

While these systems are relatively expensive to install (12,000–30,000 for an average home), annual heating costs are reduced 30%–60% and cooling costs are reduced 20%–50%. Depending on actual heating/cooling costs and heat pump size, the system can pay for itself in 8–10 years. The pump has a life expectancy of 20–25 years, while the underground piping should last 25–50 years. According to the EPA, geothermal heat pumps are the most energy-efficient, environmentally clean, cost-effective system for heating and cooling all types of buildings, including homes, offices, schools, and hospitals.

Why aren't we using more of these?

List of Key Takeaways From This Chapter

- Total fossil fuel energy sources are finite (although no one knows the exact amount of each resource remaining).
- Our society depends on a mix of different energy sources.
- Renewable energy sources are a significant portion of the total energy used in the U.S.
- All energy sources have potential benefits and potential problems.

Chapter 8 Exercises

1. Which fossil fuel contributes the most energy to the U.S.?
2. Which fossil fuel is considered the cleanest burning?
3. What is the most common use of coal in the U.S.?
4. What are two common forms of coal mining?
5. What are four hazardous substances associated with coal-burning?
6. Name two types of renewable energy.
7. What are the two major types of biofuels?
8. What are the two major problems with solar power?
9. Of the three common solar technologies, which one isn't used to produce electricity?
10. What is biomass waste?
11. What are four good points about geothermal energy?

Answers

1. Petroleum (35%)
2. Natural gas
3. Producing electricity
4. Surface mining and deep (or shaft) mining
5. SO_2, NO_x, soot, smoke, fly ash, dust, particulate matter
6. Wind, hydroelectric, wood, biofuel, solar, biomass waste, geothermal
7. Biodiesel, ethanol
8. Availability and concentration
9. Solar water-heating
10. Burning garbage, trash, etc., to produce steam and electricity
11. Clean, sustainable, abundant, reliable

CHAPTER 9

Oil in the United States

> "A century ago, petroleum – what we call oil – was just an obscure commodity; today it is almost as vital to human existence as water."
>
> James Buchan

Arguably, the two most important forms of energy in the 21st century are electricity and petroleum. In 2021, the U.S. consumed 19.78 million barrels of petroleum per day, 7.22 billion barrels in one year. About 86% was used for fuel (gasoline, diesel fuel, or jet fuel).

In 2021, about 276 million vehicles were on U.S. roads, including 107 million automobiles, 156 million trucks, 8.5 million motorcycles, and 575 thousand buses. About 99% of these vehicles use gasoline or diesel fuel and 98% were privately owned.

There were about 2.3 million electric vehicles of all types. In 2021, about 15 million new cars were sold in the United States. Electric vehicles are growing in popularity, but in 2021 sales of electric vehicles accounted for just 3.6% of total sales—about 535,000 total vehicles.

There is little doubt that for the time being, gasoline and other petroleum fuels will remain an important part of U.S. energy usage. In this chapter, we look at petroleum in a little more detail.

Learning Objectives

This chapter will help students

- Gain an understanding of the refining process
- Learn the types of molecules present in crude oil
- Have an overview of current U.S. refineries
- Understand the U.S. Strategic Petroleum Reserve
- Understand peak oil
- Understand hydraulic fracturing
- Compare the benefits and drawbacks of alcohol in gasoline

Products From One Barrel of Oil

One barrel of crude oil is 42 gallons. When refined, about 45 gallons of total products are produced. This seems strange at first glance, but volumes don't add the way that masses add. Molecules don't always arrange themselves perfectly in a liquid, and smaller molecules can fit in between larger molecules.

A classic example of this involves mixing ethanol with water. When 20 mL of ethanol mixes with 20 mL of water, we don't get 40 mL of mixture—only about 36 mL. When water and ethanol are mixed together, the molecules arrange themselves more efficiently and pack together more closely. When we separate this mixture into ethanol and water, we get 40 mL of materials.

The product distribution from a barrel of oil is controlled by the refinery, but commonly one barrel provides the following:

- 20 gallons of gasoline
- 12.5 gallons of diesel fuel (distillate)
- 3.5 gallons of jet fuel
- 1.7 gallons of hydrocarbon gas (methane, ethane, propane, and other gases)
- 0.6 gallons of residual fuel oil (fuel for ships or power plants)
- 6.3 gallons of other products

The 6.3 gallons of other products include petroleum coke, asphalt, lubricating oils, greases, waxes, and chemical feedstocks used to produce synthetic rubbers, plastics, textiles, medicines, health and beauty products, and thousands of other consumer products.

Crude oil comes in a range of colors and viscosities, from nearly colorless and clear (like water) to black and from watery to almost solid. Crude oil is classified by density [American Petroleum Institute (API) gravity] and sulfur content, with light sweet crude (low density and low sulfur content) having a higher price. This is because gasoline and diesel fuel can be produced more easily and cheaply from light sweet crude.

crude oil: unrefined liquid petroleum that is found in underground rock formations

Crude oil is not the remains of dead dinosaurs. It is the remains of sea plants and animals that lived millions of years before the dinosaurs. Over the years, layers of sand and silt covered these remains, and heat and pressure transformed them into crude oil, coal, and natural gas.

Types of Molecules Found in Crude Oil

Crude oil contains several hundred different kinds of molecules composed mainly of carbon and hydrogen (hydrocarbons). Sulfur, nitrogen, and oxygen are present in some of these molecules. The molecules are grouped into one of three classes—paraffins (alkanes), cycloparaffins (napthenes), and aromatics.

Paraffins have the general formula C_nH_{2n+2} and include compounds such as propane, butane, and pentane. Generally, paraffins range from 3 to 20 carbons and form either linear or branched molecules (Figure 9.1).

Cycloparaffins (napthenes) are ring-shaped molecules with the general formula C_nH_{2n}. The ring can be any number of carbons, although 4–8 is most typical, and extra carbon chains can be attached to the ring (Figure 9.2).

Aromatic compounds contain a benzene ring, which is a 6-carbon ring with alternating single and double bonds. There can be extra carbon chains attached to the ring in place of the hydrogen atoms, and two or more benzene rings can be "fused" together (Figure 9.3).

Sulfur, nitrogen, and oxygen atoms can replace hydrogen atoms, forming various compounds. Sulfur can be present as elemental sulfur in crude oil, or as thiols (CH_3-SH), thioethers (CH_3-S-CH_3), disulfides ($CH_3-S-S-CH_3$), and thiophenes (a 5-membered ring containing sulfur).

The nitrogen content in crude oil is normally less than 1%, with pyridine- and pyrrole-type compounds the most common (Figures 9.4 and 9.5).

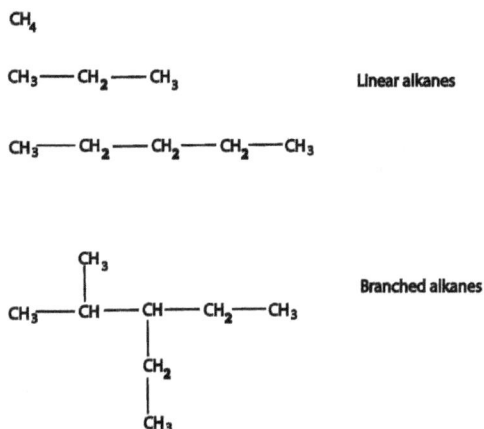

Figure 9.1   Paraffins (Alkanes)

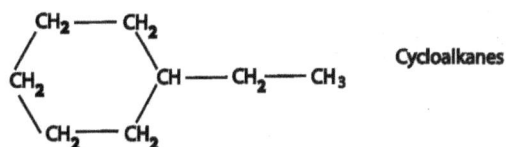

Figure 9.2   Cycloparaffins (Napthenes)

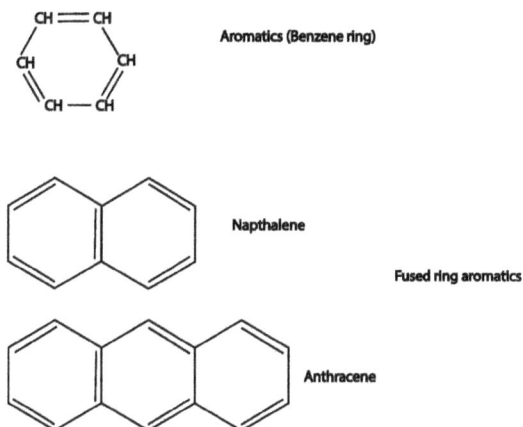

Figure 9.3   Aromatic Compounds

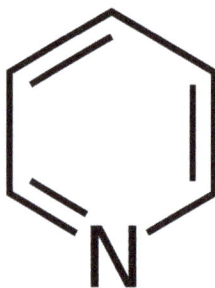

Figure 9.4   Pyridine

Calvero, "Pyrridine,"
https://commons.wikimed
ia.org/wiki/
File:Pyridine.svg, 2006.

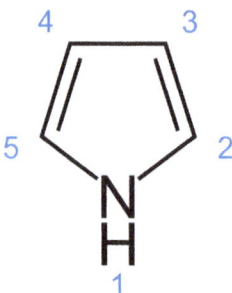

Figure 9.5   Pyrrole

Jynto, "Pyrrole,"
https://commons.wikimedia.
org/wiki/File:Pyrrole-2D-
numbered.svg, 2010.

Common types of oxygen compounds found in crude oil are summarized in Figure 9.6.

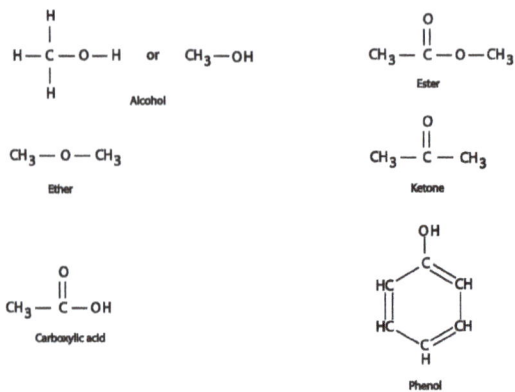

Figure 9.6   Common Types of Oxygen Compounds

Refining Process

Once the crude oil is delivered to the refinery, the oil-**refining** process begins with distillation. Crude oil is pumped through a pipe inside of a furnace, heating the oil to a high temperature. Heated oil vapors pass into a distillation tower, where the major fractions are separated by boiling temperature range (Figure 9.7).

Crude oil distillation unit and products

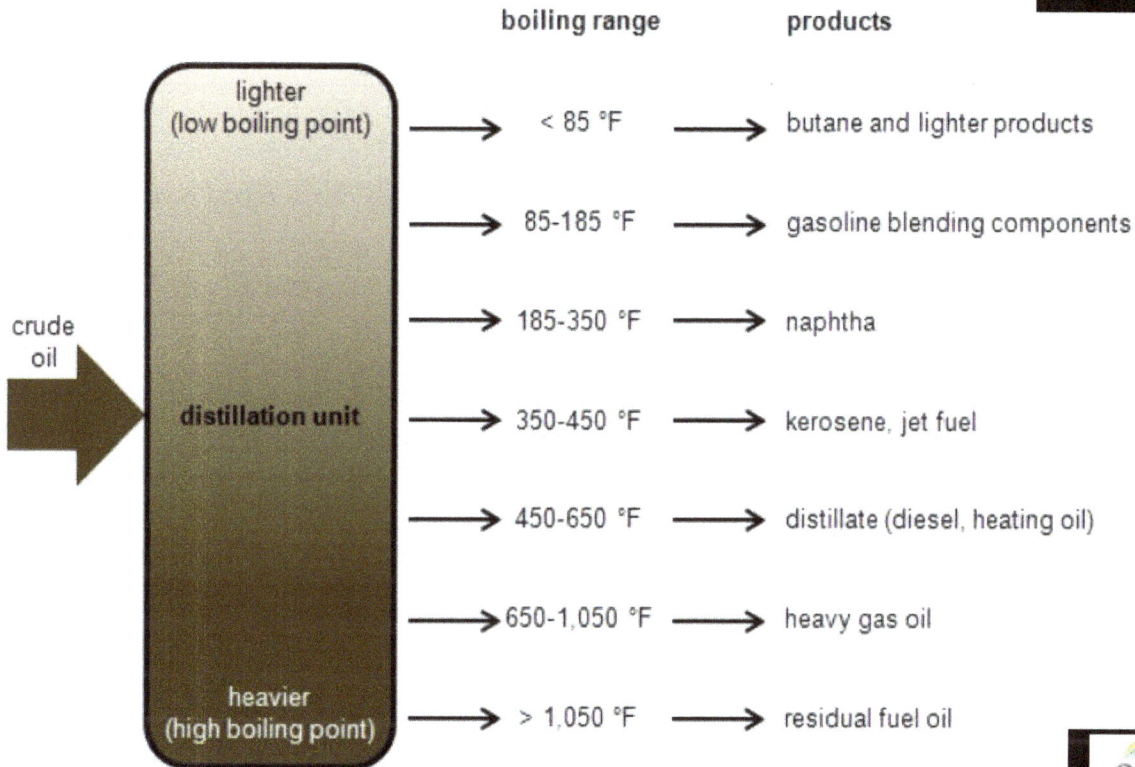

	boiling range	products
lighter (low boiling point)	< 85 °F	butane and lighter products
	85-185 °F	gasoline blending components
	185-350 °F	naphtha
distillation unit	350-450 °F	kerosene, jet fuel
	450-650 °F	distillate (diesel, heating oil)
	650-1,050 °F	heavy gas oil
heavier (high boiling point)	> 1,050 °F	residual fuel oil

crude oil →

eia

Source: U.S. Energy Information Administration.

Figure 9.7   Crude Oil Distillation

United States Energy Information Administration, https://www.eia.gov/energyexplained/oil-and-petroleum-products/refining-crude-oil-the-refining-process.php, 2021.

After distillation, the fractions are sent to one or more processing units. Major processing steps include

refining: to free something from unwanted materials or impurities

- Cracking—Large molecules are broken into smaller molecules using temperature, pressure, and catalysts.
- Alkylation—Small molecules are combined into larger molecules. Basically, this is the reverse of cracking.
- Reformation—Naptha is generally converted into gasoline components.

Other processing steps (Figure 9.8) include isomerization (converting linear molecules into branched molecules) and vacuum distillation (removing residual light molecules from the heavier oil fractions).

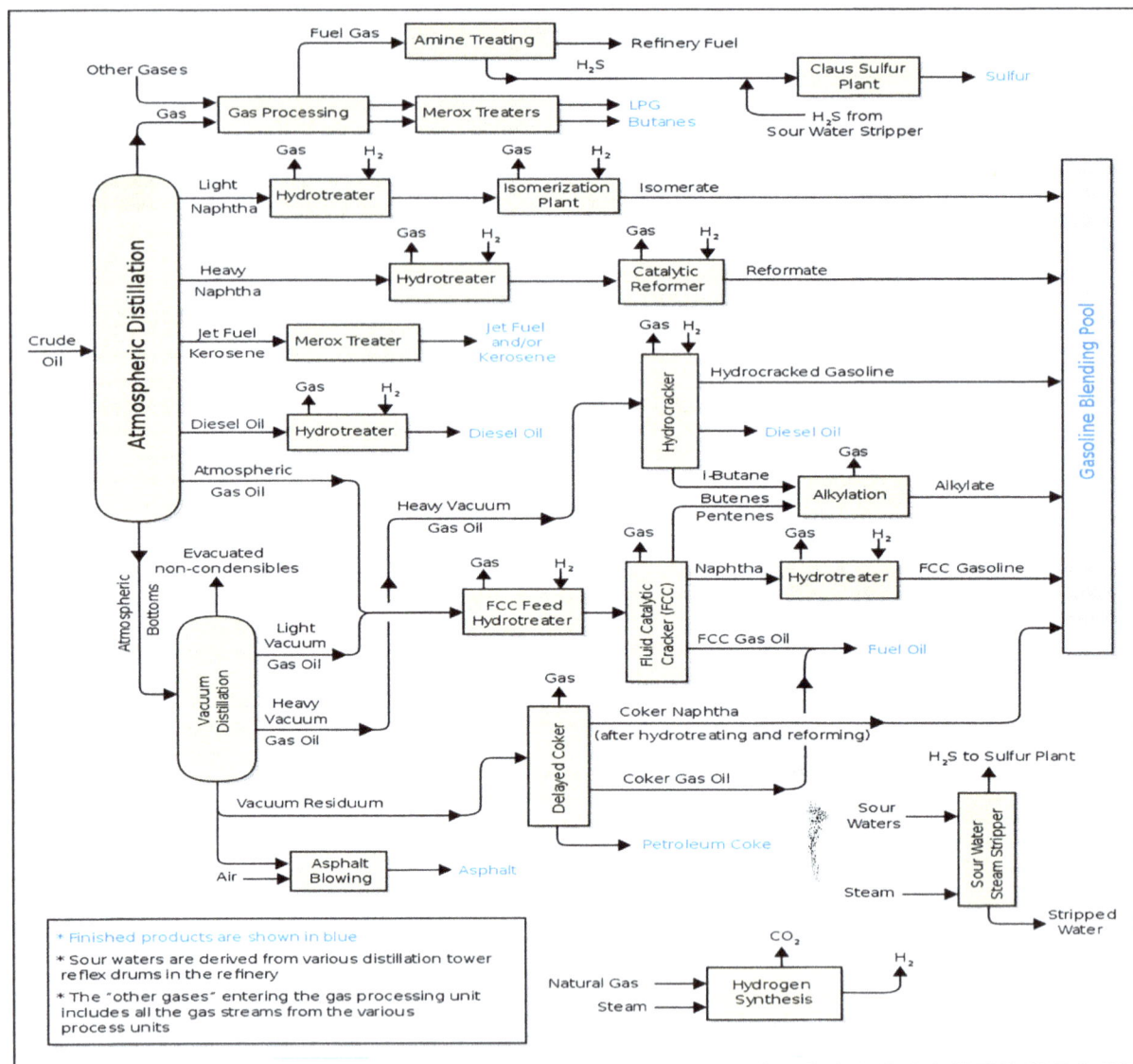

Figure 9.8   Refinery Flow Diagram

Copyright © 2018 by Begoon (CC BY-SA 3.0) at https://commons.wikimedia.org/wiki/File:RefineryFlow.svg.

Some products pass through a hydrotreater, removing sulfur and other contaminants and converting alkenes to alkanes by adding hydrogen. Other products pass through a Merox treater, removing mercaptans such as methane thiol (CH_3-SH). Final treatments include sulfuric acid treatment (removing residual alkenes, nitrogen compounds, and oxygen compounds), drying (removing residual water), and sulfur and H_2S scrubbing (removing sulfur). After cooling, the products are ready for final blending and distribution. Alcohol is added after the gasoline leaves the refinery, as are other proprietary additives.

Once crude oil reaches the refinery, it generally takes 12–24 hours to be refined into gasoline. From ground to gas station, the process takes 2–4 weeks, with much of the time spent in transit.

Current State of U.S. Refineries

Refineries are enormous industrial facilities, with a 100,000-bbl/d (barrels per day) refinery occupying about 1,500–2,000 acres. Refineries operate continuously unless shut down for repair or renovation or due to decreased demand for their products. They emit hazardous substances, including particulates, NO_x, CO, H_2S, SO_2, and small hydrocarbon molecules such as CH_4. Health hazards such as asthma, cancer, birth defects, neurological disorders, and breathing disorders are associated with living close to a refinery. Other dangers include fire, explosion, asphyxiation, and contact with corrosive or toxic substances. Make no mistake: An oil refinery is a complex and dangerous large-scale chemical plant.

As of 2021, there are 129 operating refineries in the U.S., in 30 states (Figure 9.9). Texas, Louisiana, and California have the largest number of refineries, 84 total among the three. The newest refinery, which opened in 2019 in Channelview, Texas, is a small, specialized refinery that processes 35,000 bbl/d of distillate (which is lighter than crude oil). The oldest operating refinery is the American Refinery Group at Bradford, Pennsylvania, built in 1881 and processing 11,000 bbl/d. A typical refinery processes from 100,000 to 250,000 bbl/d. The average U.S. refinery is 40 years old, and most refineries are from 50 to 120 years old. To build a 100,000-bbl/d refinery would cost about $2,500,000,000 and take 5–7 years.

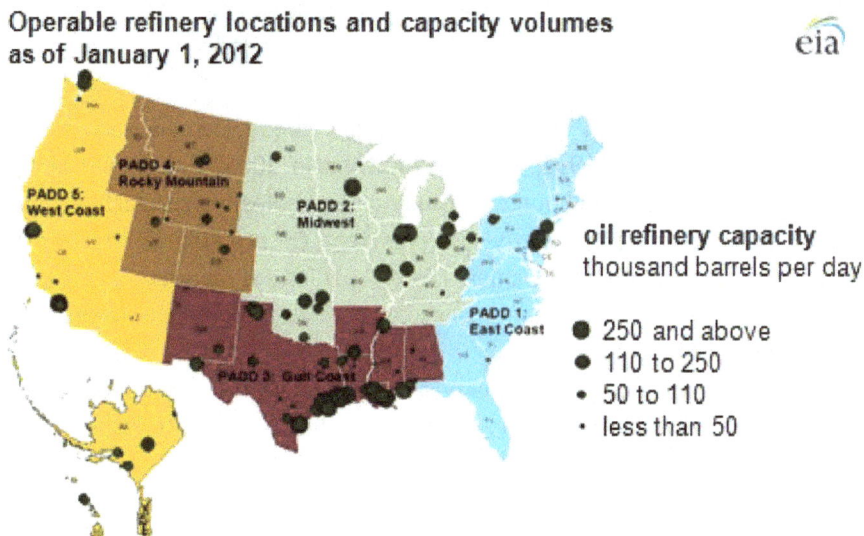

Figure 9.9   U.S. Refinery Map

United States Energy Information Administration, https://www.eia.gov/todayinenergy/detail.php?id=7170, 2012.

U.S. Strategic Petroleum Reserve

The U.S. Strategic Petroleum Reserve (SPR) was established following the 1973–1974 oil embargo in order to reduce the impact of oil disruptions on the U.S. economy. To supply the SPR, federally owned crude oil is stored in four underground salt dome caverns along the Gulf of Mexico (Figure 9.10). The total authorized capacity is 714 million barrels of oil, which at current consumption rates is 36 days' supply.

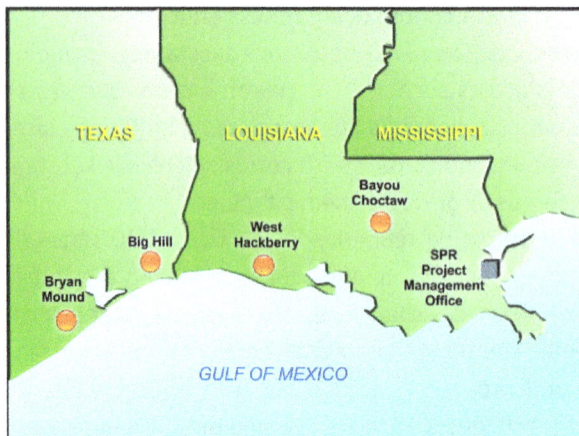

Figure 9.10   Location of SPR Storage Sites

United States Office of Energy Efficiency and Renewable Energy, https://www.energy.gov/fecm/strategic-petroleum-reserve-4, 2021.

Oil is frequently released from and replaced in the SPR. Three emergency releases have occurred since 1977: Operation Desert Storm in 1991, Hurricane Katrina in 2005, and a joint release in 2011 in response to disruptions by Libya and other countries. Crude oil test sales are conducted to ensure the SPR's readiness to respond to drawdowns ordered by the U.S. president. Oil can also be released under exchange agreements, similar to loans, with private companies. The companies are required to replace the oil in kind, with additional barrels as "interest." Nonemergency sales can be made to limit the impact of nonemergency disruptions or to raise revenues.

In June 2022, the SPR contained 498 million barrels of crude oil. It has been filled to maximum capacity once, in 2009. The maximum drawdown capacity is 4.4 million bbl/d. Once a presidential order has been given, it takes 13 days for oil from the SPR to enter the market.

Peak Oil

The total amount of recoverable crude oil is finite. Peak oil is the idea that oil production will achieve a maximum rate and then begin to decline because the oil is running out. When peak oil occurs, new crude oil discoveries won't make up for the decline in existing oil production. This is one description of peak oil. It is a very simple idea in principle, but an extremely difficult idea in practice. The major difficulty is that no one knows the total amount of recoverable crude oil available!

Peak oil has been declared several times but doesn't appear to have happened. Depending on what model you choose to use, and what assumptions you want to make, it is possible that we have already passed peak oil. Other experts forecast peak oil occurring around 2050. Still others forecast peak oil occurring in the middle of the next century.

Running out of oil is just one way for peak oil to happen. Other ways include increased expense of extracting oil, or oil alternatives becoming more economical, causing a shift away from petroleum. Depending on exactly how peak oil occurs, some will call it a catastrophe and others will call it a blessing. Either way, it will be interesting.

Fracking

Fracking, or hydraulic fracturing, is a process allowing more oil and gas to be produced from wells (Figure 9.11). This process was invented about 1947 and first used in 1949. Fracking involves pumping a mixture of water, sand, and various chemicals into the well at very high pressure. The pressure fractures the underground rock. The sand prevents the fractures from closing once the water mixture is pumped out. These fractures allow more gas and oil to flow from the well. In some cases, gas and oil wells that were abandoned have been rejuvenated by fracking.

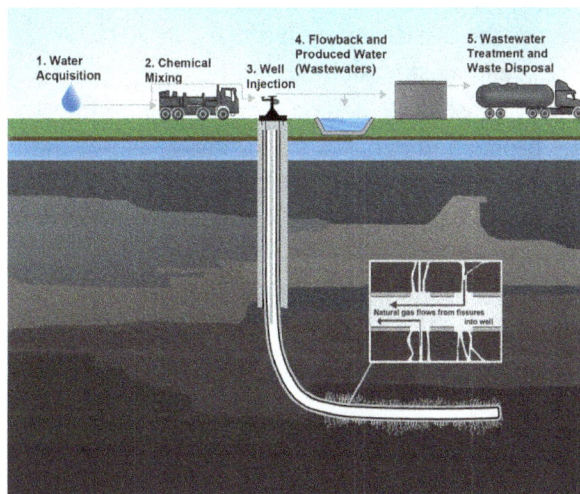

Figure 9.11   Hydraulic Fracturing (Fracking)

United States Environmental Protection Agency,
https://water.usgs.gov/owq/topics/hydraulic-fracturing/, 2016.

While the composition of fracking fluid is proprietary information, some fluids contain

- Hydrochloric or acetic acid, used to initiate fissures in the rock and clean the steel well casing
- Polyacrylamides to reduce liquid friction in the pipe
- Ethylene glycol to reduce corrosion
- Borate salts to control liquid viscosity
- Glutaraldehyde as a disinfectant

Generally, the fracking fluid is stored in containment ponds onsite, where it can be re-used if necessary.

Fracking accounts for about 95% of new oil and gas wells in the U.S., and in 2018 about half of crude oil production and two-thirds of gas production came from fracking. According to the Brookings Institution, fracking has reduced natural gas prices by 47% compared to what they would have been prior to the 2013 "fracking revolution."

However, there are significant environmental concerns. Methane emissions from fracked wells are generally higher than those from conventional wells. This doesn't seem to be a peculiarity of fracking—instead, it is due to incomplete sealing of all gas wells, allowing methane to escape into the atmosphere. Typically, 3 million–5 million gallons of water are used for the hydraulic fracturing fluid. Recycling is possible, but the fluid still must be stored until treated. Wastewater management is a significant problem, and the possibility of groundwater contamination and accidental release into the environment cannot be ignored. Hydraulic pressure can destabilize existing faults, triggering microearthquakes.

Alcohol in Gasoline

When the first U.S. oil well was dug in 1859, the oil was distilled, producing kerosene. Gasoline was discarded until the automobile industry began to grow. For a time, automobiles ran on alcohol, but they were quickly switched to gasoline because of high alcohol taxes. In 2005, Congress enacted the Renewable Fuel Standard, setting minimum requirements for the use of renewable fuels. In 2020, about 12.7 billion gallons of fuel ethanol was consumed in the U.S., and most retail motor gasoline contains 10% ethanol by volume.

There are several compelling reasons for adding ethanol to gasoline. It extends the gasoline supply; it reduces pollution, especially ozone; and it raises the octane rating of the fuel. However, fuel economy (mpg) is lower by about 3% because the total energy available per gallon of E10 (gasoline with 10% ethanol added) is lower compared to "pure" gasoline. Alcohol evaporates more quickly than gasoline, contributing to air pollution. Water has a very high affinity for alcohol. If too much water is present, phase separation can occur, producing a water–alcohol mix at the bottom and a pure gasoline layer on top. Alcohol can provide a breeding ground for microbes and fungi, although this is not a problem for ordinary consumers. It is a problem for large users and suppliers, because the microbes and fungi cause corrosion, plug filters, and reduce the fuel quality.

List of Key Takeaways From This Chapter

- A wide range of products, not just gasoline and diesel fuel, are obtained from crude oil.
- Crude oil is a complex mixture of hundreds of molecules in three general types.
- The refining process determines the amount of each kind of product available from a barrel of crude oil.
- Refineries are huge, complex chemical processing plants.
- The Strategic Petroleum Reserve acts as a buffer against disruptions in the crude oil supply.
- Peak oil can occur in several ways, not just by crude oil being completely depleted.
- Hydraulic fracturing is a method of obtaining more crude oil and gas from wells.
- The addition of alcohol to gasoline is mandated by federal law, and it has a mix of benefits and drawbacks.

Chapter 9 Exercises

1. What are three common fuels derived from crude oil?
2. Why is "light sweet" crude oil the most desirable form of petroleum?
3. What are three general types of molecules found in petroleum?
4. What is the first step in petroleum refining, and what does it accomplish?
5. What does "cracking" do?
6. What are some health hazards associated with living near a refinery?
7. What is the U.S. Strategic Petroleum Reserve?
8. What is "peak oil"?
9. What are two environmental problems associated with "fracking"?
10. Why is alcohol added to gasoline?

Answers

1. Gasoline, diesel fuel, jet fuel, residual fuel oil.
2. Gasoline can be produced easily and cheaply from light sweet crude.
3. Paraffins (alkanes), cycloparaffins (napthenes), and aromatics.
4. Distillation, which separates crude oil into fractions that can be separately processed into products.
5. Cracking uses pressure and catalysts to break large molecules into small molecules.
6. Asthma, cancer, birth defects, neurological disorders, and breathing disorders.
7. The federally owned supply of petroleum stored for emergency use.
8. When crude oil discoveries won't make up for decreased crude oil production.
9. Air pollution from methane, wastewater treatment, groundwater contamination.
10. It extends gasoline supply, reduces pollution, and raises octane rating.

CHAPTER 10

Hydrogen

> I believe that water will one day be employed as fuel, that hydrogen and oxygen which constitute it, used singly or together, will furnish an inexhaustible source of heat and light, of an intensity of which coal is not capable.
>
> Jules Verne

Hydrogen has become the gold standard for green, renewable fuels. Electricity produced from renewable sources will be used to split water into hydrogen gas and oxygen gas (H_2 and O_2). These gases will be recombined in fuel cells, generating electricity, with water as the "waste" product. No greenhouse gases will be emitted, so public health and environmental damage will no longer be a concern. Our society will achieve complete energy security, since any kind of water can be used to produce hydrogen.

It's a wonderful goal, and with a lot of good research and engineering, there is no reason why hydrogen shouldn't be part of the U.S. energy mix.

In this chapter we explore hydrogen as a fuel.

Learning Objectives

This chapter will help students

- Compare the energy content of hydrogen versus gasoline
- Gain an understanding of how hydrogen fuel cells operate
- Compare the safety of hydrogen versus gasoline
- Understand various methods of producing hydrogen
- Compare "green" hydrogen versus other types of hydrogen
- Critically evaluate hydrogen as part of the energy mixture in our society

Energy Content Versus Gasoline

Hydrogen can be used as automobile fuel. One gallon of E10 gasoline (gasoline with 10% added ethanol) contains 117,994 kJ energy. The energy density of E10 gasoline is about 42,817 kJ/kg. The energy density of hydrogen is 142,945 kJ/kg, which is about 3.3 times greater than gasoline.

However, gasoline is a liquid and hydrogen is a gas. Energy density based on mass doesn't tell the whole story. For automobiles getting 25 mpg, a 12-gallon gas tank gives a driving range of 300 miles. A 12-gallon tank of gasoline represents 1,415,928 kJ energy. The same amount of energy requires 9.905 kg hydrogen.

There are several ways to store hydrogen, but pressurized tanks are commonly used. The tank size depends on the pressure. Current applications use tank pressures of 5,000–10,000 pounds per square inch (psi). At 10,000 psi, 9.905 kg hydrogen occupies 176.7 liters (46.7 gallons) at 25 °C (77 °F). The energy density for hydrogen is 30,320 kJ/gallon compared to 117,994 kJ/gallon for gasoline—about 3.9 times smaller.

Hydrogen has higher and lower energy density values than gasoline, depending on how you compare them. Nevertheless, needing a larger tank for hydrogen isn't a big problem. Currently, hydrogen retail dispensers fill tanks in about 5 minutes, comparable to the filling time for gasoline. A fuel cell/electric motor combination is 2–3 times more efficient than a gasoline engine, so the larger tank might not be necessary if efficiency translates into higher mpg.

Other storage techniques include using sorbents (materials that absorb H_2 onto their surface), chemical hydrogen storage materials that bond hydrogen to other molecules, metal hydrides that bond hydrogen to various metal atoms, and extremely low temperature storage as either a gas or a liquid. These techniques are better suited for large-scale storage of hydrogen.

Hydrogen fuel cell

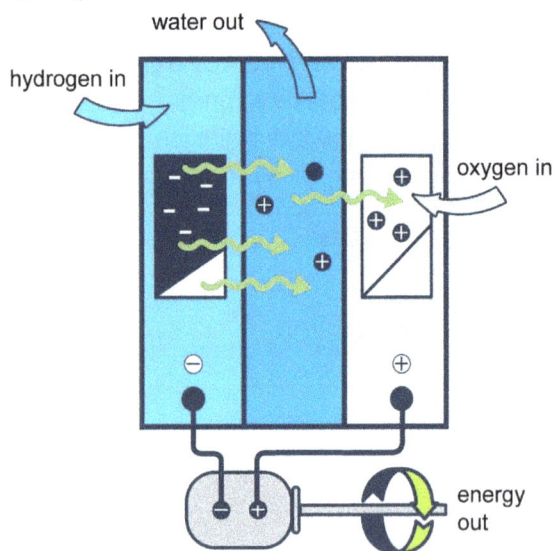

Figure 10.1   Simple Hydrogen Fuel Cell

United States Energy Information Administration, https://www.eia.gov/energyexplained/hydrogen/use-of-hydrogen.php, 2022.

Fuel Cells

Hydrogen fuel cells convert chemical energy into electrical energy by combining hydrogen and oxygen, allowing electrons to flow through a closed circuit (Figure 10.1). Hydrogen flows into the cell, where a catalyst splits hydrogen molecules into electrons and protons. The anode and cathode ("–" and "+") are separated by a polymer membrane, allowing only protons to pass through. Oxygen flows into the cathode, where it reacts with the protons and with electrons returning through the circuit, producing water. A single fuel cell typically produces 0.5 to 1.0 volts, so cells are combined to form a "stack" producing higher voltages and more power. Fuel cell systems can generate electricity at about 60% efficiency. Typical combustion plants generate electricity at about 33%–35% efficiency.

There are many types of hydrogen fuel cells, but the same electrochemical reactions occur in all of them:

$$2H_2 \; 4H+ + 4e-$$

$$O_2 + 4e- \; 2O\text{-}2$$

$$4H+ + 2O\text{-}2 \; 2H2O$$

As of June 2022, there were about 13,800 fuel cell cars and 66 fuel cell buses in operation in the U.S., almost all of them in California. California is leading the way in fuel cell vehicles, with 107 hydrogen fueling stations in operation and more planned. While fuel cell vehicles get good mileage (60–70 mpg), they are generally more expensive than conventional automobiles, due to low demand. The hydrogen refueling infrastructure is currently limited to California. Current fuel cells are not as durable as internal combustion engines, with a typical lifetime of about 75,000 miles. All these issues need to be addressed if hydrogen fuel cell vehicles are to become common.

Safety

Any chemical substance powerful enough to propel an automobile is potentially dangerous. Hydrogen is neither safer than gasoline nor less safe than gasoline. Hydrogen has different potential dangers than gasoline and substantial safety advantages over gasoline.

Hydrogen is nontoxic, and hydrogen leaks won't contaminate the environment. The only waste product produced by the fuel cell is water. Hydrogen is 14 times less dense than air, and 57 times less dense than gasoline vapor, so hydrogen tends to rise quickly into the air and disperse rapidly. Hydrocarbon vapors tend to "pool" at ground level and disperse relatively slowly. Hydrogen flames have low radiant heat, so the air around a hydrogen flame is not as hot as the air around a gasoline flame. The low radiant heat is due to water absorbing heat energy, and the absence of incandescent carbon particles radiating heat. As a result, hydrogen flames are less likely to ignite nearby combustible materials. Spontaneous ignition of hydrogen occurs at 585 °C. Gasoline spontaneously ignites at 232 °C.

The lower explosive limit (LEL) for hydrogen is 4%; that is, for hydrogen to explode in air, the hydrogen must be present at 4% by volume or higher. A hydrogen–air mixture less than 4% has too little hydrogen to explode. The LEL for gasoline is 1.2%; gasoline vapor forms explosive mixtures with air at lower concentrations than hydrogen. However, the upper explosive limit (UEL) for gasoline is 7.1%. Once gasoline vapor concentration in air exceeds 7.1%, the mixture can't explode. The UEL for hydrogen is 75%. Hydrogen has a greater range of concentrations over which a hydrogen–air mixture will explode.

Hydrogen is odorless and colorless, making it impossible to detect with human senses. The minimum ignition energy is much lower for hydrogen (0.02 mJ) than for gasoline (0.24 mJ). Hydrogen burns with an almost invisible flame, making it extremely difficult to locate. At night, it is almost impossible to detect a hydrogen flame without thermal imaging equipment. Since hydrogen flames radiate very little heat, it is possible to get too close to the flame without knowing it. Hydrogen flames give off substantial ultraviolet (UV) radiation. Some metals are subject to hydrogen embrittlement—hydrogen diffuses into the metal, weakening it. High-strength steels, titanium alloys, and aluminum alloys are vulnerable to hydrogen embrittlement.

Like any other fuel, for safe handling hydrogen requires sound engineering, proper safety procedures, and a good dose of ordinary common sense.

green: tending to preserve environmental quality (as in being recyclable, biodegradable, or nonpolluting)

hydrogen economy: an economy relying on hydrogen to provide a substantial percentage of the nation's energy needs

How Green Is My Hydrogen?

Take a minute, go back, and read the first paragraph of this chapter. I'll wait.

That paragraph describes "**green**" hydrogen, made from renewable electricity used to split water molecules. This is the goal of the "**hydrogen economy**." As of July 2022, we are nowhere close to this goal.

Currently, about 75 million metric tons (75 billion kilograms) of pure hydrogen are produced worldwide, with less than 0.1% of the total coming from electrolysis of water (according to the International Energy Agency, IEA). Oil refining and ammonia synthesis for fertilizer accounts for about 75% of world hydrogen use. In the U.S., about 95% of hydrogen comes from natural gas and less than 1% comes from green sources.

Hydrogen is typically produced by steam reformation of natural gas. Steam between 700 °C and 1,000 °C, at pressures of 40 psi to 360 psi, is combined with methane and nickel catalysts. A two-step reaction converts CH_4 into CO_2 and H_2:

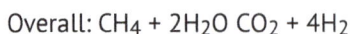

$$CH_4 + H_2O \rightarrow CO + 3H_2$$

$$CO + H_2O \rightarrow CO_2 + H_2$$

$$\text{Overall: } CH_4 + 2H_2O \rightarrow CO_2 + 4H_2$$

Producing 1.00 kg of hydrogen generates 5.46 kg of CO_2, based solely on the balanced chemical equation. This is part of the **carbon footprint** of hydrogen production. When we factor in the reformation energy (heating the materials to 700 °C–1,000 °C) and steam generation, and we account for process efficiencies (no process is 100% efficient), about 9.3 kg of CO_2 is produced for every kilogram of hydrogen. The energy content of 1 kilogram of hydrogen is 21% higher than the energy content of 1 gallon of gasoline. Combustion of 1 gallon of gasoline produces about 9.1 kg of CO_2. Clearly, there is little benefit to the environment in producing hydrogen by steam reformation of methane.

carbon footprint: the amount of greenhouse gases (CO_2 and others) produced by a particular activity

Electricity splits water into hydrogen and oxygen (Figure 10.2). Hydrogen and oxygen are produced at the cathode and anode, respectively, and are easily separated. This reaction, the electrolysis of water, was performed shortly after the first practical battery was invented in 1800.

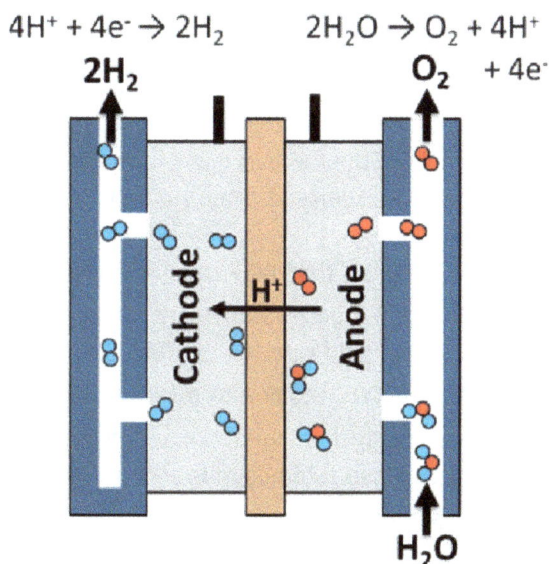

$$4H^+ + 4e^- \rightarrow 2H_2 \qquad 2H_2O \rightarrow O_2 + 4H^+$$

2H₂   **O₂**   + 4e⁻

Cathode   H⁺   Anode

H₂O

Figure 10.2   Electrolysis of Water

United States Office of Energy Efficiency and Renewable Energy, https://www.energy.gov/eere/fuelcells/hydrogen-production-electrolysis, 2022.

Steam reformation is 70%–85% efficient, while electrolysis of water is currently 70%–80% efficient (86% efficiency is projected by 2030). The carbon footprint for electrolysis depends on how the electricity is generated. Today, about 61% of U.S. electricity is produced from fossil fuels, mostly natural gas and coal. Only 20% of U.S. electricity comes from renewable sources. The *National Renewable Energy Laboratory* and the Department of Energy (NREL and DOE) find that under ideal conditions, 39 kWh of electricity is needed to produce 1 kg of hydrogen. Electrolysis of water is ~80% efficient, requiring 48 kWh of electricity to produce 1 kg of hydrogen. The carbon footprint for U.S. electricity generation is about 0.386 kg CO_2/kWh; therefore, 1 kg of hydrogen from electrolysis produces 18.5 kg CO_2. This is twice the CO_2 produced by the steam reformation process.

This is an extremely important bit of information: Until we have hydrogen produced **WITHOUT** CO_2 being released into the environment, all we are doing with hydrogen-fueled vehicles is shifting the CO_2 source away from the car's tailpipe to another location. The CO_2 from a power plant is just as damaging as CO_2 from an automobile. As of July 2022, driving a hydrogen-powered car isn't reducing total CO_2, and the overall benefit to the environment is difficult to assess. Carbon dioxide isn't the whole story. Other emissions from gasoline engines are nitrogen oxides (NO_x), particulates, carbon-containing compounds, carbon monoxide, and formaldehyde. None of these are present in exhaust from hydrogen fuel cell vehicles. Conversely, fossil fuel power plants also produce the same mix of emissions as gasoline engines.

Many other methods of hydrogen production are being researched, including production by bacteria and algae; using concentrated solar energy to decompose water into hydrogen and oxygen (the HydroSol-2 plant in Spain, for example); and various thermochemical reactions to split water, using heat in place of electricity. Microwaving ground plastics with iron oxide and aluminum oxide has been shown to recover 97% of the hydrogen in the plastic. All these methods require much more work before going into production scale.

Color is used to describe hydrogen production, based upon source and how CO_2 is handled (Table 10.1). Purple, pink, and red hydrogen involve nuclear power, introducing a whole new set of social, political, environmental, and technical problems. Turquoise hydrogen is interesting, and is an attractive option depending on how the high temperatures are achieved. Gray and brown/black aren't really an option, for the reasons given earlier.

Blue hydrogen is based on the premise that CO_2 can be captured (which is relatively easy) and permanently sequestered/stored. Biological sequestration stores carbon in grasslands, forests, soil, and the ocean. Arguably this is the best method for sequestration. Storing CO_2 in living organisms is certainly not permanent. Geological sequestration injects CO_2 into porous rocks in underground formations. Technological sequestration stores carbon using innovative technologies. Some of the more interesting technologies being developed are those to convert CO_2 into organic compounds such as methane, or to produce graphene—a form of carbon used in electronics. All sequestration methods have potential short-term and long-term issues.

That leaves green hydrogen produced by electrolysis of water, using electricity produced by renewable resources, as the only guaranteed environmentally safe method of producing hydrogen, having zero CO_2 emissions.

Table 10.1. Colors of Hydrogen

Color	Description
Green	Produced by water electrolysis using renewable electricity
Blue	Produced from fossil fuels, with CO_2 captured and stored underground; classified as "carbon neutral" since no CO_2 is emitted
Gray	Produced by steam reformation of methane, with CO_2 eventually released into the air
Brown (or black)	Produced by coal gasification; both CO_2 and carbon monoxide are produced and released into the air
Turquoise	Thermal splitting of methane, producing solid carbon and hydrogen; still in experimental development
Purple	Using nuclear power and heat to split water by chemo-thermal electrolysis
Pink	Electrolysis of water, using electricity produced by nuclear power
Red	Produced by high-temperature catalytic splitting of water using a nuclear produced thermal energy
White	Naturally occurring hydrogen

Future of the Hydrogen Economy

The hydrogen economy is based on the idea that hydrogen will provide a substantial portion of the nation's energy needs. But "substantial portion" isn't particularly informative. How much is *substantial?*

Hydrogen production using renewable energy sources can be an important energy storage technique. Renewable energy sources like solar and wind have built-in limitations—they provide energy only when the sun is shining or the wind is blowing. Storing electricity, as electricity, is extremely difficult. Converting electrical energy into chemical energy, producing chemical compounds that can be readily stored and used when needed, is much more practical.

There is also a scale problem that must be addressed. Given the amount of energy used in the U.S., can electrolytic hydrogen act as a replacement source? Let's look at a couple of scenarios.

Scenario 1: We want to replace all gasoline-powered automobiles with hydrogen-fueled automobiles, using electrolysis of water to obtain the hydrogen.

In 2021, the U.S. used 3,879 billion kWh of electricity (according to the EIA). This is 3.879×10^{12} kWh.

In 2021, the U.S. consumed 134.83 billion gallons of E10 gasoline (according to the EIA). One gallon provides 117,944 kJ, so total gasoline consumption accounted for 1.590×10^{16} kJ of energy.

In 2021, the average fuel economy of U.S. automobiles was 36 mpg. A fuel cell/electric motor car gets up to 70 mpg, about twice the mileage of gasoline.

One kilogram of hydrogen provides 142,945 kJ of energy. To replace gasoline with hydrogen requires 1.112×10^{11} kg of hydrogen. Increased fuel efficiency requires only half of this hydrogen, 5.560×10^{10} kg.

It takes 48 kWh to produce 1 kg of hydrogen. To produce 5.560×10^{10} kg of hydrogen requires 2.669×10^{12} kWh of electricity. In 2021, this is two-thirds of the total electricity production of the U.S. If we use the theoretical minimum value of 39 kWh/1 kg hydrogen, it would require 2.168×10^{12} kWh, more than half of U.S. electrical production.

To fulfill the conditions of this scenario requires 6.548×10^{12} kWh of electrical capacity for hydrogen fuel and other electricity needs, 68% more than is currently installed. There are about 11,000 utility-scale (1 MW or larger) power plants in the U.S. We would need ~7,500 more. Ideally, none of these additional power plants would be fossil-fueled.

How would this change our carbon footprint?

In 2021, there were 289.5 million automobiles. The EPA estimates 4,600 kg of CO_2 per car. This is 1.332×10^{12} kg CO_2.

The carbon footprint of U.S. electricity is currently 0.386 kg CO_2/kWh. Total CO_2 for electricity generation in 2021 was about 1.497×10^{12} kg. Combined CO_2 from electricity and automobiles is 2.829×10^{12} kg in 2021.

If we had an electrical capacity of 6.548×10^{12} kWh, allowing us electricity for hydrogen production and other uses, and if the 7,500 new power plants were green or blue, then the carbon footprint would be determined by CO_2 emitted from the existing fossil fuel power plants, 1.497×10^{12} kg. This would be a 47% reduction in carbon footprint. If we kept the same relative mix of fossil fuel to renewable fuel power plants, the carbon footprint for 6.548×10^{12} kWh would be 2.528×10^{12} kg, an 11% reduction.

In any case, I don't see Scenario 1 happening anytime soon.

Scenario 2: We want to replace fossil fuels for electricity generation with hydrogen, reducing our carbon footprint.

Around 61% of the 3.879×10^{12} kWh of electricity produced in the U.S. comes from fossil fuels. This is 8.518×10^{15} kJ. Energy is energy, regardless of the source, and in this scenario there is no mileage advantage to consider. Replacing fossil fuel–produced electricity requires 5.959×10^{10} kg of hydrogen. Obtaining this hydrogen would be a significant challenge. Worldwide production of pure hydrogen is about 7.5×10^{10} kg. About 75% of it comes from natural gas, which produces CO_2. Less than 0.1% of global dedicated production comes from electrolysis (according to the IEA). An enormous increase in green or blue hydrogen production would be needed.

Hydrogen should play a role in our society, but I don't believe electrolytic hydrogen is "the answer" to our energy needs. To achieve the hydrogen economy, green or blue methods of hydrogen production must be developed that can produce the tremendous quantities of hydrogen needed to replace the fossil fuels currently used. Whether or not these hydrogen production methods can be developed remains to be seen.

List of Key Takeaways From This Chapter

- Hydrogen contains more energy/kilogram than gasoline.
- Fuel cells using hydrogen produce electricity more efficiently than does ordinary combustion.
- The only emission from hydrogen fuel cells is water.
- Hydrogen is not more dangerous than gasoline. Hydrogen and gasoline each have their own safety issues.
- The vast majority of hydrogen produced in the U.S. and the world is from steam reformation of methane.
- Less than 1% of hydrogen in the U.S. and less than 0.1% of worldwide hydrogen production is "green."
- Currently, the carbon footprint to produce 1 kg hydrogen is larger than that from burning 1 gallon of gasoline.
- The hydrogen economy is a worthwhile goal, but achieving it will require that we address significant engineering and societal challenges.

Chapter 10 Exercises

1. Why is hydrogen considered a very desirable fuel?
2. Generally, how does a hydrogen "fuel cell" work?
3. Is hydrogen safer to use than gasoline?
4. What is "green" hydrogen?
5. What is meant by "hydrogen economy"?
6. What are the most common industrial uses of hydrogen?
7. What is the most common method of producing hydrogen gas?
8. What does the expression "carbon footprint" mean?

Answers

1. The only waste product produced by burning hydrogen is water.
2. Hydrogen and oxygen combine to produce water and electricity.
3. Neither hydrogen nor gasoline is inherently safer than the other. Both substances require proper equipment and procedures for safe handling.
4. Green hydrogen is made from renewable electricity.
5. The hydrogen economy relies on hydrogen providing a substantial percentage of the nation's energy requirements.
6. Refining petroleum and manufacturing fertilizer.
7. Steam reformation of methane.
8. Amount of greenhouse gases (especially CO_2) produced by a given activity.

CHAPTER 11

Electrochemistry

> Each metal has a certain power, which is different from metal to metal, of setting the electric fluid in motion.
>
> Alessandro Volta

Electricity is critically important in our society, with about 38% of all energy usage in the United States coming from electricity. Electric usage is expected to increase by nearly 30% over the next 30 years. Without electricity, our entire modern society would collapse.

Entire books have been written about the history of electricity and batteries. In this chapter, I focus on the relationships between chemical reactions and electricity.

Learning Objectives

This chapter will help students

- Use oxidation rules to assign oxidation numbers to elements
- Identify oxidizing/reducing agents
- Balance oxidation–reduction reactions using the half-reaction method
- Calculate the voltage produced by oxidation–reduction reaction
- Understand the operation of modern batteries

Invention of the Battery

As part of a debate between Luigi Galvani (1737–1798) and Alessandro Volta (1745–1827) concerning the origin of electricity in animals, Volta constructed the first modern battery in 1799 (Figure 11.1). Volta's battery was made from silver and zinc coins, with salt water–soaked paper or cloth separating the coins. Eventually, other metal combinations were used to build batteries, producing different amounts of electricity.

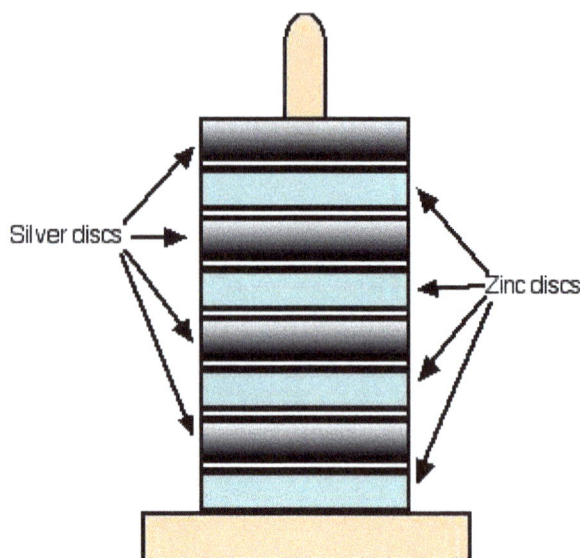

Silver discs

Zinc discs

Figure 11.1   First Modern Battery

The battery was a world-changing invention. Before the battery, scientists used static electricity in their experiments. The problem with static electricity is that it lasts just a fraction of a second. Longer experiments with electricity were impossible. The battery provided a steady electricity supply lasting several hours. A few months after Volta's battery design was published, scientists used batteries to decompose water, salt, and other substances into their chemical elements. Many of these elements, such as sodium and chlorine, had never been seen as pure substances.

While batteries were important tools in physics and chemistry laboratories, they had little impact outside the laboratory because there were no electrical or electronic devices requiring them. Batteries and the devices using them developed side by side.

Oxidation Numbers

Two of the earliest electrical scientists were Humphry Davy (1778–1829) and his assistant Michael Faraday (1791–1867). Among Davy's contributions were the isolation of sodium, potassium, barium, calcium, strontium, and magnesium by passing electricity through molten salts. This was the first time these elements were obtained in their pure, elemental form.

Faraday's experiments indicated that a fundamental relationship exists between the amount of electrical current and the mass of a given element. Faraday popularized the usage of the terms **electrode**, **ion**, **anion**, and **cation**. Eventually, Faraday developed the idea of **oxidation numbers** to help keep track of the movement of electrons between elements during a chemical reaction.

When we assign oxidation numbers, we pretend that the atoms in a compound are ions—*but we are just pretending!* Most compounds are **NOT** composed of ions, but we can imagine them that way. There are specific rules for assigning oxidation numbers to atoms:

1. The oxidation number of an element, in its elemental form, is zero (0).
2. The oxidation number of a simple ion (a single atom with "+" or "–" charge) is equal to the charge: size and sign!
3. In compounds, the oxidation number of hydrogen is almost always +1.
4. In compounds, the oxidation number of oxygen is almost always –2.
5. In compounds, the oxidation number of elements in the first column of the periodic table (Group 1A) is +1.

electrode: a conductor through which electricity enters or leaves an object or a substance

ion: an atom or a molecule with an electric charge, either + or –

anion: an atom or a molecule with a negative electric charge

cation: an atom or a molecule with a positive electric charge

oxidation number: the charge that the atom would have in a compound, *if the compound were composed of ions*

6. In compounds, the oxidation number of elements in the second column of the periodic table (Group 2A) is +2.
7. In compounds, the oxidation number of elements in the 17th column of the periodic table (Group 7A) is almost always −1.
8. In a neutral molecule (no net electric charge), the sum of "+" oxidation numbers and "−" oxidation numbers = 0.
9. In ions (a net electric charge, either + or −), the sum of "+" oxidation numbers and "−" oxidation numbers equals the charge on the ion.

Let's apply these rules in some examples:

A. CH_4. The oxidation number of hydrogen in compounds is +1 (Rule 3 above). The molecule is neutral: no net electric charge. The oxidation number of carbon (C) must be −4. Here's a simple way to figure out the oxidation numbers: List all the atoms shown in the chemical formula and fill in the known oxidation numbers from the rules given. You can then determine what the other oxidation numbers should be. For our example,

C	−4
H	+1
H	+1
H	+1
H	+1
Net charge	0

B. Fe_2O_3. The oxidation number of oxygen is −2 (Rule 4). With three oxygens, the sum of "−" oxidation numbers is −6. Since no net charge is shown for the molecule, there must be a total +6 to balance the −6 oxidation number. This +6 must be assigned to the Fe atoms, +3 to each. Therefore, the oxidation number of Fe is +3.
C. NaOH. The oxidation number of Na is +1 (Rule 5). The oxidation number of O is −2 (Rule 4), and the oxidation number of H is +1 (Rule 3).
D. MnO_4^-. The oxidation number of O is −2 (Rule 4). Since there is a charge (we have an ion), Rule 9 applies. Four oxygen atoms gives a total of −8 oxidation numbers. One of these oxidation numbers accounts for the charge; therefore, Mn must be +7.

oxidation: loss of electrons by an atom, molecule, or ion during a chemical reaction
reduction: gain of electrons by an atom, molecule, or ion during a chemical reaction

Oxidation numbers are a bookkeeping tool allowing us to track the movement of electrons during chemical reactions. Moving electrons produce electric current (electricity). When electrons are lost, **oxidation** occurs; when electrons are gained, **reduction** occurs.

Oxidizing/Reducing Agents

Oxidation and reduction occur in many chemical reactions. Here's a simple example:

$$Zn + 2HCl \rightarrow ZnCl_2 + H_2$$

Assigning oxidation numbers, we see that Zn went from 0 to +2 (loss of electrons is oxidation), H went from +1 to 0 (gain of electrons is reduction), and Cl didn't change (started with −1 and remained −1).

Oxidizing agents cause some other material to become oxidized. **Reducing agents** cause some other material to become reduced. The oxidizing agent MUST contain the element that accepts the electrons, while the reducing agent MUST contain the element that donates the electrons. We could rewrite the above equation into separate oxidation and reduction steps:

Oxidation step	$Zn \rightarrow Zn^{+2} + 2e^-$
	$2HCl \rightarrow 2H^+ + 2Cl^-$
Reduction step	$2H^+ + 2e^- \rightarrow H_2$

oxidizing agent: a substance that causes oxidation by gaining electrons and being reduced

reducing agent: a substance that causes reduction by losing electrons and being oxidized

In the oxidation step, zinc lost 2 electrons. Loss of electrons is oxidation. The electrons can't just disappear out of the universe, they must go somewhere. The electrons go to the hydrogen ions (H^+). The 2 hydrogen ions gained 2 electrons from zinc, causing zinc to become oxidized. Similarly, zinc lost 2 electrons, allowing the 2 hydrogen ions to become reduced.

In our redox reaction, zinc is the reducing agent; it gave 2 electrons to the hydrogen ions, causing them to be reduced. HCl is the oxidizing agent. **NOTICE:** Oxidizing and reducing agents are always reactants (on the left-hand side of the reaction arrow). Even though only one element gains or loses electrons, the **AGENT** is always the *substance* containing the element. The oxidizing agent is HCl, not just H^+.

The best way to identify oxidizing and reducing agents is to look for elements whose oxidation number changes in the chemical equation. If the oxidation number becomes more positive (the element is being oxidized), then the substance containing this element is the *reducing agent*. If the oxidation number becomes more negative, then the substance containing this element is the *oxidizing agent*. Consider this example:

$$MnO_4^- + C_2O_4^{-2} \rightarrow Mn^{+2} + CO_2$$

Assigning oxidation numbers, we see that Mn goes from +7 to +2. Since Mn gained electrons, it was reduced; therefore, MnO_4^- is the oxidizing agent. Oxygen (O) didn't change; it remained −2 throughout the reaction. Carbon (C) went from +3 to +4. Since C lost electrons, it was oxidized, and therefore $C_2O_4^{-2}$ is the reducing agent.

If you look carefully at the reaction, you see that it is not balanced. We have 8 oxygen atoms on the left side, but only 2 on the right. We have 2 carbon atoms on the left side, but only 1 on the right. If we tried to balance this reaction using the method shown in Chapter 6, we would have a very, very difficult time. Instead, we balance redox reactions using the half-reaction method.

Balancing Reactions by the Half-Reaction Method

The half-reaction method involves balancing the oxidation and reduction reactions separately. Once the two half-reactions are balanced, they are recombined.

The procedure followed for each half-reaction is this:

1. Balance all elements, other than hydrogen and oxygen, using the method described in Chapter 6.
2. Balance oxygen by adding water molecules: For every oxygen atom needed, one water molecule is added.
3. Balance hydrogen by adding hydrogen ion (H^+): For every hydrogen atom needed, one hydrogen ion is added.
4. Compare the total charge on the left side and the right side. If the charges are different, add electrons (e^-) to equalize the charges.

In our example, the oxidation half-reaction is this:

$$C_2O_4^{-2} \ CO_2$$

First, balance the carbon atoms by putting a coefficient of 2 in front of CO_2:

$$C_2O_4^{-2} \ 2CO_2$$

Carbon is balanced. Notice that we have the same number of oxygen atoms on each side of the reaction. We also have the same number of hydrogen atoms on each side of the reaction (!).

The total charge must be the same on both sides. We have "–2" charge on the left, and "0" charge on the right, so add 2 electrons to the right side to balance the charge:

$$C_2O_4^{-2} \ 2CO_2 + 2e^-$$

This is a balanced oxidation half-reaction; we have the same number and kinds of atoms, and the same total charge, on both sides of the equation. Balancing this half-reaction was easy; sometimes we need to do more work.

The reduction half-reaction is

$$MnO_4^- \ Mn^{+2}$$

We have the same number of manganese atoms on both sides, so manganese is balanced. To balance oxygen, we add one water molecule for every missing oxygen atom. The 4 oxygen atoms require 4 water molecules.

$$MnO_4^- \ Mn^{+2} + 4H_2O$$

Oxygen is balanced. For every missing hydrogen atom, we add one hydrogen ion (H^+). The 8 hydrogen atoms from 4 water molecules require 8 hydrogen ions.

$$8H^+ + MnO_4^- \ Mn^{+2} + 4H_2O$$

On both sides, we have the same number and kinds of atoms. There is a net charge of +7 for the reactants (eight "+" charges from hydrogen ions, one "–" charge from MnO_4^-), and a net charge of +2 for the products (from Mn^{+2}). We add 5 electrons to the reactant side of the equation:

$$8H^+ + MnO_4^- + 5e^- \; Mn^{+2} + 4H_2O$$

This is a balanced reduction half-reaction. Notice that in an oxidation half-reaction, electrons are products, while in a reduction half-reaction, electrons are reactants.

Now combine the half-reactions. To combine reactions, the same number of electrons must be produced as are reacted. Since 5 electrons are reacted in the reduction reaction, and 2 electrons are produced in the oxidation reaction, multiply the reduction half-reaction by 2, and the oxidation half-reaction by 5, to give the following two reactions:

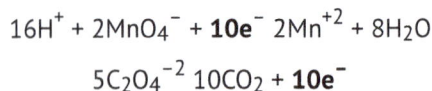

$$16H^+ + 2MnO_4^- + \mathbf{10e^-} \; 2Mn^{+2} + 8H_2O$$

$$5C_2O_4^{-2} \; 10CO_2 + \mathbf{10e^-}$$

Add the two reactions by adding all reactants and all products in the two half-reactions. Cancel out substances that are identical on both sides of the equation:

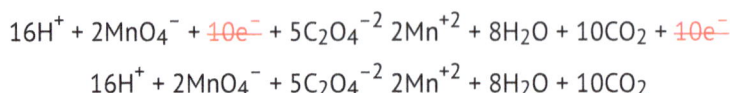

$$16H^+ + 2MnO_4^- + \cancel{10e^-} + 5C_2O_4^{-2} \; 2Mn^{+2} + 8H_2O + 10CO_2 + \cancel{10e^-}$$

$$16H^+ + 2MnO_4^- + 5C_2O_4^{-2} \; 2Mn^{+2} + 8H_2O + 10CO_2$$

Finally, check the equation to ensure that mass and charge balance have been achieved:

Reactants	Products
16 H	16 H
2 Mn	2 Mn
28 O	28 O
10 C	10 C
+4	+4

This is an example of balancing a redox reaction, by the half-reaction method, *in acid solution*. In acid solution, a hydrogen ion is present.

We can also balance the reaction *in basic solution*. First, balance the reaction in acid solution. Then neutralize the hydrogen ions by adding an equal number of hydroxide ions to both sides of the reaction:

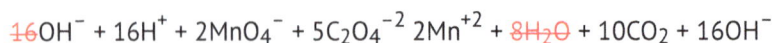

$$\cancel{16}OH^- + 16H^+ + 2MnO_4^- + 5C_2O_4^{-2} \; 2Mn^{+2} + \cancel{8H_2O} + 10CO_2 + 16OH^-$$

Combine hydroxide ions with hydrogen ions to form water molecules, and cancel out as many water molecules as are common on both sides of the equation.

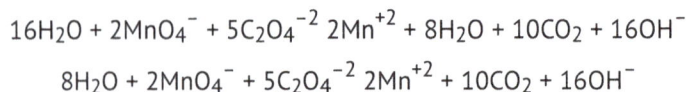

$$16H_2O + 2MnO_4^- + 5C_2O_4^{-2} \; 2Mn^{+2} + 8H_2O + 10CO_2 + 16OH^-$$

$$8H_2O + 2MnO_4^- + 5C_2O_4^{-2} \; 2Mn^{+2} + 10CO_2 + 16OH^-$$

Check to ensure that mass and charge balance:

Reactants	Products
16 H	16 H
2 Mn	2 Mn
36 O	36 O
10 C	10 C
−12	−12

Voltage Calculations

Oxidation and reduction must occur simultaneously: We can't have reduction without oxidation. Every oxidation or reduction reaction has a specific **voltage** associated with it. Voltages are affected by concentration of substances, temperature, and other factors, so generally voltages are given for standard conditions. Standard conditions are typically 25 °C, 1 atm pressure (if gases are involved), and 1.00 mole/liter concentration for solutions, and they are measured relative to a standard hydrogen electrode (SHE). The voltage of a SHE is defined as 0.000 volts, which serves as a reference value for all other voltages.

voltage: the pressure from an electrical circuit's power source that pushes electrons through the circuit, enabling the electrons to do work

For example, the oxidation of zinc metal is

$$Zn \rightarrow Zn^{+2} + 2e^-$$

If I connect a zinc electrode to a SHE, under standard conditions, and measure the voltage produced, I get a value of +0.762 volts. This is the standard oxidation potential (SOP) of zinc.

Generally, reactions are tabulated as standard reduction potentials (SRP). The difference between the SOP and the SRP is just a sign change: The SRP for zinc is −0.762 volt, corresponding to

$$Zn^{+2} + 2e^- \rightarrow Zn$$

Let's combine our zinc electrode with a silver electrode, recreating Volta's first battery. The reduction of silver is

$$Ag^+ + e^- \rightarrow Ag$$

The SRP for silver is 0.799 volt. The overall redox reaction is

$$2Ag^+ + Zn \rightarrow Zn^{+2} + 2Ag$$

Zinc is oxidized. Silver is reduced.

The voltage for the complete redox reaction (called E_{cell}) is given by the formula

$$E_{cell} = SRP_{\text{reduction step}} - SRP_{\text{oxidation step}}$$

$$E_{cell} = SRP_{\text{silver}} - SRP_{\text{zinc}}$$

$$E_{cell} = 0.799 \text{ volt} - (-0.762 \text{ volt})$$

$$E_{\text{cell}} = 1.561 \text{ volts}$$

Please note that when half-reaction voltages are given as SRP, we always subtract the SRP for the oxidation half-reaction from the SRP of the reduction half-reaction. In the balanced reaction we have $2Ag^+$, but the voltage for silver reduction is not doubled!

If E_{cell} is positive, then the cell is producing electric current. If E_{cell} is negative, there are two possibilities: First, we may have written the redox reaction backward. Second, we may be deliberately running the reactions in reverse, by supplying electric current.

Voltage and **current** are the electricity equivalents of potential energy (voltage) and kinetic energy (current). The "cell" we are talking about can also be called a galvanic cell—or, more commonly, a battery.

Batteries

current: the amount of electricity flowing in a circuit; the flow of electrons through a circuit

electrolyte: a liquid solution containing ions that is electrically conducting through the movement of those ions but does not conduct electrons

Batteries are commonly divided into wet-cell and dry-cell types. The wet-cell battery was the original rechargeable battery. In a wet-cell battery, a liquid **electrolyte** such as sulfuric acid is present. Metal plates are immersed in the acid. The battery must remain upright to prevent spillage. Hydrogen gas is produced during operation and must be vented. Probably the most common wet-cell battery is the lead–acid storage battery used in automobiles, trucks, boats, and other vehicles (Figure 11.2).

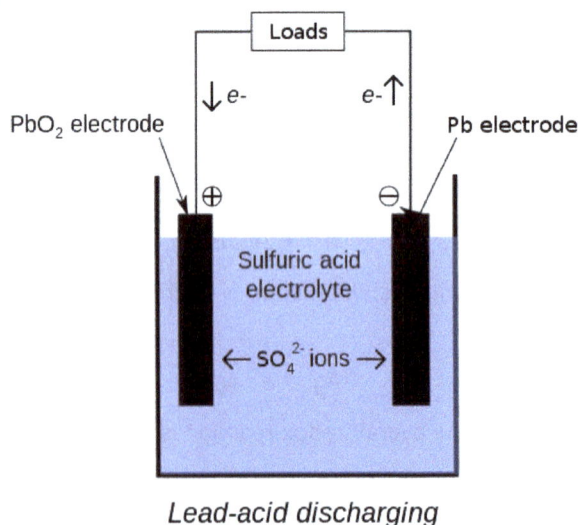

Lead-acid discharging

Figure 11.2   Lead–Acid Battery (One Cell)

Copyright © 2007 by BatteryGuy (CC BY-SA 3.0) at
https://en.wikipedia.org/wiki/File:Lead-acid_discharging.svg.

During use, the oxidation reaction occurs at the lead electrode:

$$Pb + SO_4^{-2} \text{ } PbSO_4 + 2e^-$$

Reduction occurs at the PbO_2 electrode:

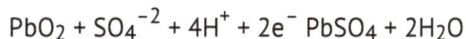

$$Pb O_2 + SO_4^{-2} + 4H^+ + 2e^- \; Pb SO_4 + 2H_2O$$

The overall reaction is

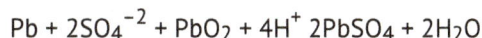

$$Pb + 2SO_4^{-2} + PbO_2 + 4H^+ \; 2Pb SO_4 + 2H_2O$$

A six-cell lead–acid battery produces about 2.1 volts/cell, or about 12.6 volts total.

Dry-cell batteries don't use a pool of liquid electrolyte. Instead, a paste electrolyte is used that contains a minimum amount of liquid. This paste doesn't spill, so the battery's position isn't important. Alkaline batteries are one common type of dry-cell battery (Figure 11.3).

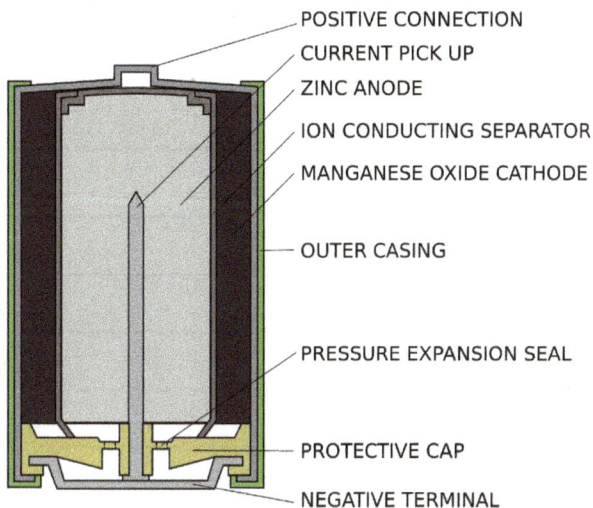

Figure 11.3   Alkaline Battery

Tympanus, https://commons.wikimedia.org/wiki/File:Alkaline-battery-english.svg, 2011.

The paste electrolyte is a mixture of potassium hydroxide (KOH) with water. This paste is mixed with powdered zinc to make the anode, where oxidation occurs. Manganese dioxide (MnO_2) mixed with coal dust makes the cathode, where reduction occurs.

The oxidation reaction is

$$Zn + 2OH^- \; ZnO + H_2O + 2e^-$$

The reduction reaction is

$$2MnO_2 + H_2O + 2e^- \; Mn_2O_3 + 2OH^-$$

Overall,

$$Zn + 2MnO_2 \; ZnO + Mn_2O_3$$

Common alkaline batteries come in AAA, AA, C, and D sizes. Size A and B exist, but they aren't used in the U.S. All alkaline batteries have the same voltage: 1.5 V. Rectangular 9-volt batteries are essentially six AAAA batteries wired in series. The different sizes contain different amounts of chemicals and produce different amounts of current.

The energy available in a battery is affected by its voltage and **amp-hours**. Total joules of energy is found from

$$\text{joules} = V \times \text{amp} - \text{hours} \times 3,600$$

amp-hours: the amount of energy charge in a battery that enables 1 ampere of current to flow for 1 hour

Table 11.1 shows typical values for alkaline and automobile batteries. For comparison, a 90-mph fastball has about 162 joules kinetic energy. A 4,400-lb truck (2,000 kg) traveling 89 mph (40 m/s) has a kinetic energy of 1,600,000 joules ($E_k = \frac{1}{2} mv^2$). Batteries contain relatively large quantities of energy. Even a AAA battery contains about 50% more energy than the average rifle bullet. Like any other energy source, batteries should be treated with the proper level of caution. If alkaline batteries are misused, leakage of corrosive chemicals (KOH), overheating, or explosion is possible.

Table 11.1. Energy Content of Various Batteries

Size	Volts	Amp-hours	Joules
AAA	1.5	1.15	6,210
AA	1.5	2.85	15,390
C	1.5	7.8	42,120
D	1.5	15	81,000
9-volt	9	0.57	18,468
"Car"	12	48	2,073,600

There are two general types of lithium battery: primary lithium batteries (lithium metal batteries) and lithium-ion batteries. Primary lithium batteries are not rechargeable and are disposable. They produce from 1.5 to 3.7 volts, and are commonly small, button-shaped batteries. They are used in flashlights, toys, cameras, watches, and remote controls and as power sources for motherboards in laptop computers. Lithium metal is the anode, and a metal oxide like MnO_2 is the cathode. Common lithium metal batteries are "button" batteries, such as the CR2032.

Lithium-ion batteries are rechargeable. They are used in all sorts of electronic devices, including cell phones, laptops, tablets, power tools, and electric cars. Lithium-ion technology is rapidly evolving, and the specific chemistry, performance, cost, and safety vary across the various types.

One common set of reactions is this:

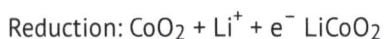

$$\text{Oxidation: } LiC_6 \; C_6 + Li^+ + e^-$$

$$\text{Reduction: } CoO_2 + Li^+ + e^- \; LiCoO_2$$

During recharging, the two reactions are reversed, with Li^+ moving into C_6.

Recharging time depends on the battery size. For electronic devices, typically 2 to 3 hours are required. During use, the lithium moves from anode to cathode, providing the current. The electrolyte that is used to carry Li^+ is a lithium salt ($LiPF_6$, for example), in an organic solvent (ethylene carbonate, $C_3H_4O_3$). Alternately, solid electrolytes such as ceramics or polymers are used in place of the liquid electrolyte.

Lithium-ion batteries have good energy/weight ratios (100–265 Wh/kg vs. ~45 Wh/kg for a lead–acid battery) and relatively high voltage (~3.7 volts), and they last for ~0.9 amp-hours. They are reliable, relatively easy to recharge, do not contain toxic metals such as lead or cadmium, and are relatively lightweight and compact.

However, they can overheat, and they can be damaged by high voltages. When they are used in automobiles, recharging the battery can take from 20 minutes to 50 hours, depending on the type of charging station used and the battery size. They have lower energy density than gasoline (~12,700 Wh/kg). Currently, they have a 10-year to 20-year lifetime. Replacement cost varies but is somewhere between 2,000 and 10,000. If the battery is overcharged or damaged, it can short-circuit and start a fire.

According to data compiled by the National Transportation Safety Board (NTSB), hybrid vehicles (combined gasoline and electric) had 3,474 fires/100,000 vehicles sold. Gasoline-powered vehicles had 1,530 fires/100,000 vehicles sold, while electric vehicles had 25 fire/100,000 vehicles sold. However, lithium battery fires tend to burn hotter, last longer, and require much more water to extinguish than do gasoline fires. The two most common causes of electric vehicle fire are (1) damage to the battery, caused by a collision, and (2) fires that occur while recharging.

List of Key Takeaways From This Chapter

- The invention of batteries allowed scientists to explore the relationship between electricity and chemistry.
- The use of oxidation numbers allows us to follow the flow of electrons between substances during chemical reactions.
- The flow of electrons produces electric current.
- Different combinations of materials can be used to make batteries that have different voltages.
- All batteries are dangerous if mishandled.

Chapter 11 Exercises

Part A. Assign oxidation numbers to the individual elements in the following compounds.

1. H_2S
2. BaO
3. P_2O_5
4. $Cr_2O_7^{-2}$
5. CH_2O
6. HBr
7. CO
8. H_2SO_4

Part B. In the following chemical reactions, identify which substance is being oxidized and which substance is being reduced.

1. $Fe + NiCl_2$ $FeCl_2 + Ni$
2. $K + HNO_3$ $KNO_3 + H_2$
3. $CH_4 + 2O_2$ $CO_2 + 2H_2O$
4. $C_2H_4 + H_2$ C_2H_6
5. $H_2SO_4 + 2NaOH$ $2Na^+ + SO_4^{-2} + 2H_2O$

Part C. Use the half-reaction method and balance the following equations under the listed conditions.

1. $CH_3OH + MnO_4^-$ $CO_2 + H_2O + MnO_2$ (basic solution)
2. $Cr_2O_7^{-2} + HNO_2$ $Cr^{+3} + NO_3^-$ (acidic solution)
3. $TeO_3 + N_2O_4$ $Te + NO_3^-$ (acidic solution)
4. As $H_2AsO_4^- + AsH_3$ (acidic solution)
5. $MnO_4^- + S_2O_3^{-2}$ $S_4O_6^{-2} + Mn^{+2}$ (acidic solution)
6. $PH_3 + I_2$ $H_3PO_2^- + I^-$ (acidic solution)

Part D. For each of the electrochemical reactions shown below, identify the oxidation reaction and the reduction reaction. Calculate the total cell voltage (E_{cell}) = E_{red} − E_{oxid}. Finally, tell if the reaction is spontaneous as written or if it has been reversed.

1. $3Ag^+(aq) + Fe^0(s)$ $Fe^{+3}(aq) + 3Ag^0(s)$
2. $5K^0(s) + MnO_4^- + 8H^+$ $5K^+(aq) + Mn^{+2}(aq) + 4H_2O$
3. $3Cu^{+2}(aq) + 2Fe^0(s)$ $2Fe^{+3}(aq) + 3Cu^0(s)$
4. $2Al^{+3}(aq) + 6Cl^-(aq)$ $2Al^0(s) + 3Cl_2(aq)$
5. $Au^+(aq) + Li(s)$ $Li^+(aq) + Au^0(s)$

The following standard reduction potentials are needed to calculate E_{cell}.

$Ag^+(aq) + e^- \; Ag^0(s)$	$E^0 = 0.799$ volt
$Fe^{+3}(aq) + 3e^- \; Fe^0(s)$	$E^0 = 0.331$ volt
$K^+(aq) + e^- \; K^0_{\cdot}(s)$	$E^0 = -2.936$ volts
$MnO_4^- + 8H^+ + 5e^- \; Mn^{+2}(aq) + 4H_2O$	$E^0 = 1.507$ volts
$Cu^{+2}(aq) + 2e^- \; Cu^0(s)$	$E^0 = 0.339$ volt
$Al^{+3}(aq) + 3e^- \; Al^0(s)$	$E^0 = -1.677$ volts
$Cl_2(aq) + 2e^- \; 2Cl^-(aq)$	$E^0 = 1.396$ volts
$Au^+(aq) + e^- \; Au^0(s)$	$E^0 = 1.690$ volts
$Li^+(aq) + e^- \; Li^0(s)$	$E^0 = -3.040$ volts

Part E. Answer the following questions briefly but clearly.

1. What are two naturally occurring forms of electricity?
2. What is an "ion"?
3. What is the difference between a "cation" and an "anion"?
4. What is oxidation? What is reduction?
5. What is the difference between voltage and current?

Answers

Part A.

1. The oxidation number of hydrogen is +1. Since there are 2 hydrogen atoms in the compound, and no net electrical charge shown, the oxidation number of sulfur must be −2.

    $$+1 \quad -2$$

    $$H_2S$$

2. The oxidation number of oxygen is −2. With no net electrical charge shown, the oxidation number of barium is +2. [Note: Barium is a Group IIA (2) element, and by Rule 3 would have an oxidation number of +2.]
3. The oxidation number of oxygen is −2. There is no net electrical charge shown. Five oxygen atoms result in a total negative oxidation state of −10. This must be exactly balanced by the positive oxidation state of phosphorous. With two phosphorous atoms, we assign an oxidation number of +5 to each phosphorous.
4. The oxidation number of each oxygen is −2. For 7 oxygen atoms, this accounts for a total of −14. There is a −2 charge shown for the ion, and so two of the −14 oxidation numbers can be assigned to the charge. This leaves −12, which must be balanced by the positive oxidation state of chromium. With 2 chromium atoms in the ion, we assign each an oxidation number of +6.
5. The oxidation number of oxygen is −2. The oxidation number of hydrogen is +1. Since the total positive oxidation state (2 × +1) exactly balances the total negative oxidation state, the oxidation state of carbon must be zero (0). NOTE: While the oxidation state of elements in their elemental form must be zero, it is also possible for an element to have a zero oxidation state when combined in a compound.
6. The oxidation number of hydrogen is +1, while that of bromine is −1.
7. The oxidation state of oxygen is −2; therefore, carbon must be +2.
8. The oxidation state of hydrogen is +1. The oxidation state of oxygen is −2. The result with 4 oxygens is a total of −8, while 2 hydrogens produce a total of +2. With no net charge shown, the oxidation state of sulfur must be +6.

Part B.

1. Elemental iron has an oxidation number of 0. The iron in iron (II) chloride has an oxidation number of +2. The nickel in nickel (II) chloride has an oxidation number of +2. The oxidation number of elemental nickel is 0. The oxidation number of chloride didn't change. Therefore, iron has been oxidized (it is the reducing agent), while nickel ion has been reduced.
2. Elemental potassium has an oxidation number of 0. In nitric acid (HNO_3), the oxidation number of hydrogen is +1 and the oxidation number of oxygen is −2, so the oxidation number of nitrogen is +5. In potassium nitrate, potassium is +1; the other elements didn't change oxidation number (N = +5, O = −2). The oxidation number of hydrogen in elemental hydrogen (H_2) is zero (0). Therefore, potassium has been oxidized and hydrogen has been reduced.

3. The oxidation number of carbon in methane (CH_4) is −4, while its oxidation number in carbon dioxide is +4. The oxidation number of hydrogen in all compounds is +1. The oxidation number of oxygen in its elemental form is 0, while its value in carbon dioxide and water is −2. Therefore, carbon has been oxidized, and methane is the reducing agent. Oxygen has been reduced, and it is the oxidation agent.

4. The oxidation number of carbon in ethene (C_2H_4) is −2, while in ethane (C_2H_6) carbon has an oxidation number of −3. The oxidation number of hydrogen in ethene and ethane is +1. The oxidation number of elemental hydrogen is 0. Therefore, carbon in ethene has been reduced, and ethene is the oxidizing agent. Elemental hydrogen has been oxidized, and it is the reducing agent.

5. No element changes its oxidation number in this reaction. Therefore, nothing has been oxidized, and nothing has been reduced. This is not a redox reaction.

Part C.

1. $CH_3OH + MnO_4^-$ $CO_2 + H_2O + MnO_2$ (basic solution)
    a. CH_3OH CO_2 **oxidation reaction**
    b. $CH_3OH + H_2O$ CO_2 balance O with water
    c. $CH_3OH + H_2O$ $CO_2 + 6H^+$ balance H with H^+
    d. $CH_3OH + H_2O$ $CO_2 + 6H^+ + 6e^-$ balance charge with e^-
    e. MnO_4^- MnO_2 **reduction reaction**
    f. MnO_4^- $MnO_2 + 2H_2O$ balance O with water
    g. $4H^+ + MnO_4^-$ $MnO_2 + 2H_2O$ balance H with H^+
    h. $4H^+ + MnO_4^- + 3e^-$ $MnO_2 + 2H_2O$ balance charge with e^-
    i. Multiply "h" × 2, multiply "d" × 1:

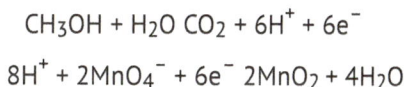

    $$CH_3OH + H_2O\ CO_2 + 6H^+ + 6e^-$$

    $$8H^+ + 2MnO_4^- + 6e^-\ 2MnO_2 + 4H_2O$$

    j. Add two equations together:

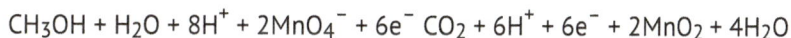

    $$CH_3OH + H_2O + 8H^+ + 2MnO_4^- + 6e^-\ CO_2 + 6H^+ + 6e^- + 2MnO_2 + 4H_2O$$

    k. Cancel common substances on both sides:

    $$CH_3OH + \cancel{H_2O} + \cancel{8}\,2H^+ + 2MnO_4^- + \cancel{6e^-}\ CO_2 + \cancel{6H^+} + \cancel{6e^-} + 2MnO_2 + \cancel{4}\,3H_2O$$

    (Note: Only $6H^+$ can be canceled, so $2H^+$ are left as reactants. Similarly, one water is canceled from the reactants, leaving three waters in the products.)

    l. Add OH^- to both sides to neutralize H^+:

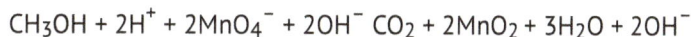

    $$CH_3OH + 2H^+ + 2MnO_4^- + 2OH^-\ CO_2 + 2MnO_2 + 3H_2O + 2OH^-$$

    m. Combine H^+ with OH^-, cancel out waters:

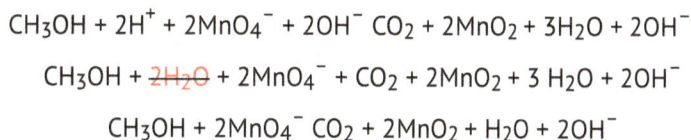

    $$CH_3OH + 2H^+ + 2MnO_4^- + 2OH^-\ CO_2 + 2MnO_2 + 3H_2O + 2OH^-$$

    $$CH_3OH + \cancel{2H_2O} + 2MnO_4^- + CO_2 + 2MnO_2 + 3\,H_2O + 2OH^-$$

    $$CH_3OH + 2MnO_4^-\ CO_2 + 2MnO_2 + H_2O + 2OH^-$$

n. Check reactants/products for mass/charge balance:

	Reactants	Products
C	1	1
H	4	4
O	9	9
Mn	2	2
Charge	-2	-2

2. $Cr_2O_7^{-2} + HNO_2 \rightarrow Cr^{+3} + NO_3^-$ (acidic solution)

    a. $Cr_2O_7^{-2} \rightarrow Cr^{+3}$ **reduction reaction**

    b. $Cr_2O_7^{-2} \rightarrow 2Cr^{+3}$ balance Cr

    c. $Cr_2O_7^{-2} \rightarrow 2Cr^{+3} + 7H_2O$ balance O with water

    d. $Cr_2O_7^{-2} + 14H^+ \rightarrow 2Cr^{+3} + 7H_2O$ balance H with H^+

    e. $Cr_2O_7^{-2} + 14H^+ + 6e^- \rightarrow 2Cr^{+3} + 7H_2O$ balance charge with e^-

    f. $HNO_2 \rightarrow NO_3^-$ **oxidation reaction**

    g. $H_2O + HNO_2 \rightarrow NO_3^-$ balance O with water

    h. $H_2O + HNO_2 \rightarrow NO_3^- + 3H^+$ balance H with H^+

    i. $H_2O + HNO_2 \rightarrow NO_3^- + 3H^+ + 2e^-$ balance charge with e^-

    j. Multiply "i" × 3:

        $3H_2O + 3HNO_2 \rightarrow 3NO_3^- + 9H^+ + 6e^-$

    k. Add "e" and "j" together:

        $Cr_2O_7^{-2} + 14H^+ + 6e^- + 3H_2O + 3HNO_2 \rightarrow 2Cr^{+3} + 7H_2O + 3NO_3^- + 9H^+ + 6e^-$

    l. Cancel common substances on both sides of equation:

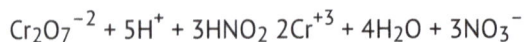

        $Cr_2O_7^{-2} + \cancel{14}\,5H^+ + \cancel{6e^-} + \cancel{3H_2O} + 3HNO_2 \rightarrow 2Cr^{+3} + \cancel{7}\,4H_2O + 3NO_3^- + \cancel{9H^+} + \cancel{6e^-}$

        $Cr_2O_7^{-2} + 5H^+ + 3HNO_2 \rightarrow 2Cr^{+3} + 4H_2O + 3NO_3^-$

m. Check reactants/products for mass/charge balance:

	Reactants	Products
Cr	2	2
O	13	13
H	8	8
N	3	3
Charge	+3	+3

3. $TeO_3 + N_2O_4$ Te + NO_3^- (acidic solution)

   a. TeO_3 Te **reduction reaction**

   b. TeO_3 Te + $3H_2O$ balance O with water

   c. $6H^+ + TeO_3$ Te + $3H_2O$ balance H with H^+

   d. $6e^- + 6H^+ + TeO_3$ Te + $3H_2O$ balance charge with e^-

   e. N_2O_4 NO_3^- **oxidation reaction**

   f. N_2O_4 $2NO_3^-$ balance N normally

   g. $2H_2O + N_2O_4$ $2NO_3^-$ balance O with water

   h. $2H_2O + N_2O_4$ $2NO_3^- + 4H^+$ balance H with H^+

   i. $2H_2O + N_2O_4$ $2NO_3^- + 4H^+ + 2e^-$ balance charge with e^-

   j. Multiply "h" by 3:

   $6H_2O + 3N_2O_4$ $6NO_3^- + 12H^+ + 6e^-$

   k. Add two equations together:

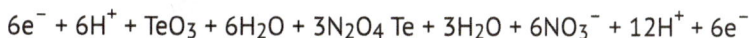

   $6e^- + 6H^+ + TeO_3 + 6H_2O + 3N_2O_4$ Te + $3H_2O + 6NO_3^- + 12H^+ + 6e^-$

   l. Cancel out common substances on both sides of equation:

   $\cancel{6e^-} + \cancel{6H^+} + TeO_3 + \cancel{6}\ 3H_2O + 3N_2O_4$ Te + $\cancel{3H_2O} + 6NO_3^- + \cancel{12}\ 6H^+ + \cancel{6e^-}$

   $TeO_3 + 3H_2O + 3N_2O_4$ Te + $6NO_3^- + 6H^+$

m. Check reactants/products for mass/charge balance:

	Reactants	Products
Te	1	1
O	18	18
H	6	6
N	6	6
Charge	0	0

4. As $H_2AsO_4^-$ + AsH_3 (basic solution)

   a. As $H_2AsO_4^-$ **oxidation reaction**
   b. $4H_2O$ + As $H_2AsO_4^-$ balance O with water
   c. $4H_2O$ + As $H_2AsO_4^-$ + $6H^+$ balance H with H^+
   d. $4H_2O$ + As $H_2AsO_4^-$ + $6H^+$ + $5e^-$ balance charge with e^-
   e. As AsH_3 **reduction reaction**
   f. $3H^+$ + As AsH_3 balance H with H^+
   g. $3e^-$ + $3H^+$ + As AsH_3 balance charge with e^-
   h. Multiply "d" × 3; multiply "g" × 5:

   $12H_2O$ + 3As $3H_2AsO^-$ + $18H^+$ + $15e^-$

   $15e^-$ + $15H^+$ + 5As $5AsH_3$

   i. Add two equations together:

   $12H_2O$ + 3As + $15e^-$ + $15H^+$ + 5As $3H_2AsO_4^-$ + $18H^+$ + $15e^-$ + $5AsH_3$

   j. Cancel out common substances on both sides:

   $12H_2O$ + 8As + ~~$15e^-$~~ + ~~$15H^+$~~ $3H_2AsO_4^-$ + ~~18~~ $3H^+$ + ~~$15e^-$~~ + $5AsH_3$

   $12H_2O$ + 8As $3H_2AsO_4^-$ + $3H^+$ + $5AsH_3$

   k. Check reactants/products for mass/charge balance:

	Reactants	Products
H	24	24
O	12	12
As	8	8
Charge	0	0

5. $MnO_4^- + S_2O_3^{-2}$ $S_4O_6^{-2} + Mn^{+2}$ (acidic solution)
   a. MnO_4^- Mn^{+2} **reduction reaction**
   b. MnO_4^- $Mn^{+2} + 4H_2O$ balance O with water
   c. $8H^+ + MnO_4^-$ $Mn^{+2} + 4H_2O$ balance H with H^+
   d. $5e^- + 8H^+ + MnO_4^-$ $Mn^{+2} + 4H_2O$ balance charge with e^-
   e. $S_2O_3^{-2}$ $S_4O_6^{-2}$ **oxidation reaction**
   f. $2S_2O_3^{-2}$ $S_4O_6^{-2}$ balance S normally
   g. $2S_2O_3^{-2}$ $S_4O_6^{-2} + 2e^-$ balance charge with e^-
   h. Multiply "d" × 2; multiply "f" × 5:

   $$10e^- + 16H^+ + 2MnO_4^- \quad 2Mn^{+2} + 8H_2O$$

   $$10S_2O_3^{-2} \quad 5S_4O_6^{-2} + 10e^-$$

   i. Add two reactions together, cancel common substances on both sides:

   $$\cancel{10e^-} + 16H^+ + 2MnO_4^- + 10S_2O_3^{-2} \quad 2Mn^{+2} + 8H_2O + 5S_4O_6^{-2} + \cancel{10e^-}$$

   $$16H^+ + 2MnO_4^- + 10S_2O_3^{-2} \quad 2Mn^{+2} + 8H_2O + 5S_4O_6^{-2}$$

   j. Check reactants/products for mass/charge balance:

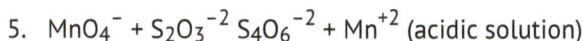

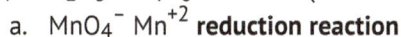

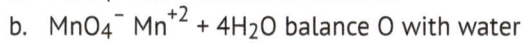

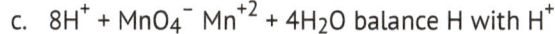

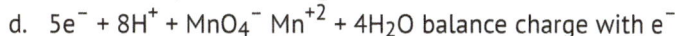

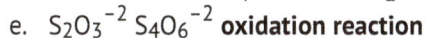

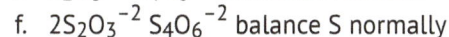

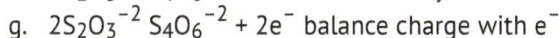

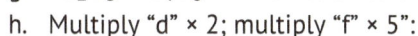

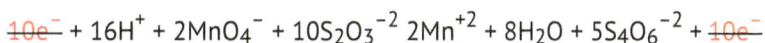

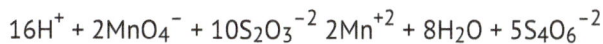

	Reactants	Products
H	16	16
Mn	2	2
O	38	38
S	20	20
Charge	−6	−6

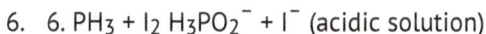

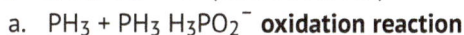

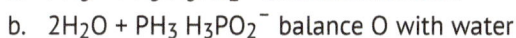

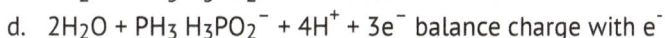

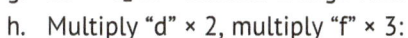

6. 6. $PH_3 + I_2$ $H_3PO_2^- + I^-$ (acidic solution)
   a. $PH_3 + PH_3$ $H_3PO_2^-$ **oxidation reaction**
   b. $2H_2O + PH_3$ $H_3PO_2^-$ balance O with water
   c. $2H_2O + PH_3$ $H_3PO_2^- + 4H^+$ balance H with H+
   d. $2H_2O + PH_3$ $H_3PO_2^- + 4H^+ + 3e^-$ balance charge with e^-
   e. I_2 I^- **reduction reaction**
   f. I_2 $2I^-$ balance I normally
   g. $2e^- + I_2$ $2I^-$ balance charge with e^-
   h. Multiply "d" × 2, multiply "f" × 3:

   $$4H_2O + 2PH_3 \quad 2H_3PO_2^- + 8H^+ + 6e^-$$

   $$6e^- + 3I_2 \quad 6I^-$$

i. Add reactions, cancel out common substances:

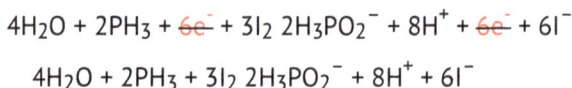

$$4H_2O + 2PH_3 + \cancel{6e^-} + 3I_2 \ 2H_3PO_2^- + 8H^+ + \cancel{6e^-} + 6I^-$$

$$4H_2O + 2PH_3 + 3I_2 \ 2H_3PO_2^- + 8H^+ + 6I^-$$

j. Check reactants/products for mass/charge balance:

	Reactants	Products
H	14	14
O	4	4
P	2	2
I	6	6
Charge	0	0

Part D.
1. E_{cell} = 0.799 V – 0.331 V = 0.468 V Spontaneous
2. E_{cell} = 1.507 V – (–2.936 V) = 4.443 V Spontaneous
3. E_{cell} = 0.339 V – 0.331 V = 0.008 V Spontaneous
4. E_{cell} = –1.677 V – 1.396 V = –3.073 V Nonspontaneous
5. E_{cell} = 1.690 V – (–3.040 V) = 4.730 V Spontaneous

Part E.
1. Static and lightning.
2. A small molecule or a single atom with an electrical charge.
3. A cation is an ion with a positive electric charge. An anion is an ion with a negative charge.
4. Oxidation is loss of electrons by an atom or a molecule. Reduction is gain of electrons by an atom or a molecule.
5. The voltage is the "pressure" that pushes electrons through a circuit, whereas the current is the amount of electricity flowing through the circuit.

www.ingramcontent.com/pod-product-compliance
Lightning Source LLC
Chambersburg PA
CBHW081539220326
41598CB00036B/6492